The Economy of Death

Books by Richard J. Barnet

The Economy of Death   1969

Intervention and Revolution   1968

After Twenty Years (WITH MARCUS RASKIN)   1965

Security in Disarmament (EDITED WITH

RICHARD A. FALK)   1965

Who Wants Disarmament?   1960

Richard J. Barnet

THE ECONOMY
OF DEATH

Atheneum

NEW YORK

1969

TO THE MEMORY OF

James P. Warburg

WHO UNDERSTOOD MANY YEARS AGO THAT

FOREIGN POLICY BEGINS AT HOME

I HAVE SET BEFORE YOU LIFE AND DEATH, BLESSING
AND CURSE; THEREFORE CHOOSE LIFE THAT YOU
AND YOUR DESCENDANTS MAY LIVE.

Deuteronomy, 30:19

Contents

Contents

The Economy of Death

INTRODUCTION

What Is Security?

S ince 1946 the taxpayers have been asked to contribute more than one trillion dollars for national security. Each year the federal government spends more
than 70 cents of every budget dollar on past, present,
and future wars. The American people are devoting
more resources to the war machine than is spent by all
federal, state, and local governments on health and hospitals, education, old-age and retirement benefits, public assistance and relief, unemployment and social
security, housing and community development, and the
support of agriculture. Out of every tax dollar there is
about 11 cents left to build American society.

Nations, like families, reveal themselves through
budgets. No personal document tells more about a
man's values or his hopes and fears than the family
budget. Similarly, the way to size up a nation is to
examine the national budget. But the real cost of America's search for security through armaments cannot be
adequately measured in money. Ordinary mortals, even
rich mortals and the Congressmen who vote the appropriations, cannot understand what a trillion dollars is.
To comprehend the magnitude of our investment in
killing power, we need to look at what we have sacri-
ficed for it. The Economy of Life in America has been
starved to feed the Economy of Death.

The result of this gigantic investment in security has

been to make the American people among the most insecure on the planet. Perhaps the most important index of this fact is that Americans are more afraid to walk the streets of Washington than the streets of Saigon. These feelings are a direct reflection of an increasingly violent society. College presidents are beginning to administer their universities with the aid of armed helicopters. A carnival of anger has swept across the ghettos of every American metropolis; the fragile nature of our civilization is attested by the burned and looted stores of the inner city. Teachers, garbage men, and taxi drivers have taken turns in bringing America's largest city to standstill. The President of the United States no longer dares to travel among the people.

An incipient civil war has already begun. Young against old. Suburb against inner city. Black against white. According to the President's Violence Commission, an American is four times more likely to be a victim of violence than a West European. In his inaugural address President Nixon alluded to the crisis of spirit in the United States and counseled people not to shout at one another. But good manners, important as it is to restore them, are no answer to the American crisis. As a college president recently observed on the day he resigned in the wake of a student rebellion, "A society that does not correct its own ills cannot expect peace."

The pattern is tragically familiar. The world historian Arnold Toynbee, who has traced the fate of the great empires of the past, finds that most have gone down to defeat not by invasion from without but because of social dissolution within. Mighty nations that do not respond to the needs of their own people have traditionally tried to solve problems and overcome frustrations through violence abroad and repression at home. In the process, they have hastened their own exit from center stage. The greatest security problems for a nation are the hostility and frustration of its own citizens.

In the struggle for survival, social institutions, like living things, achieve security only if they are able to adapt to changing conditions. Besides the increasing public unhappiness due to the failure to invest in American society, there are other conditions, also largely ignored, which threaten our security. One is the international political environment. As one observer has put it, the American people are living under an indeterminate death sentence. We have heard so much scare talk about nuclear war that we discount it. Yet statistically the odds increase year by year. More weapons of greater destructive power in more hands in more conflicts represent a growing security threat. The fact that we can't measure the threat or predict when a nuclear bomb will go off doesn't mean that we can safely assume that none ever will. Making more bombs, the instinctive response of our government to danger from abroad, only increases the risk. Since 1945 the institution of war has been incompatible with the survival of the human race. Yet the United States, the largest military power in the world, the dominant force in the world economy, and the annual consumer of 60 percent of the world's consumable resources, has responded to this threat by making war preparation the central activity of its government. In so doing, it has set the tone for the whole international community.

The systematic destruction of the natural environment threatens a slower but surer death for American civilization. Every six months in the United States we eliminate forest areas the size of the state of Rhode

Island. Every city with 500,000 people dumps 50 million gallons of sewage a year into our streams, rivers, and lakes. Pesticides and pollution have resulted in the extermination of Delaware shad and Merrimack shellfish and have made Lake Erie unfit for fish of any kind. About 50 percent of all the space in our cities is taken up by automobiles, roads, parking lots, and gas stations. The automobile is one of the chief causes of air pollution that is costing a steadily increasing number of lives each year. As an ecologist recently testified before Congress, we are playing Russian roulette with the oxygen supply. We are tampering with the delicate ecological balance that supports human life. This is literally borrowing against the lives of our children. Moreover, the world is depleting mineral resources and fresh water so fast that the Secretary-General of the United Nations warns, "The future of life on earth could be endangered." Yet these mammoth security problems merit exactly 1.8 percent of the federal budget. That is what we spend on all federal environmental programs.

No less important a security problem is the waste of human resources. The investment in the Economy of Death contributes to this waste in two ways. First, our brightest, most creative people are given the incentive to apply their talents to the essentially insoluble problem of finding security in the technology of death. As a result, the society can afford to support only a tiny group to think about rescuing the environment, solving the problem of mass transportation, or figuring out how to rebuild the cities. Second, the budget of the Department of Defense has gobbled up resources that otherwise could have been made available for human investment. Education budgets generally fall as defense budgets rise. The cost to the nation of its failure to solve the education problem is incalculable. Every glassy-eyed child turned out of a boring school is a social loss. In health, too, the failure to invest undermines national security. The infant mortality rate in the United States is higher than in Hungary or Lebanon. The reason, of course, is that in many rural areas and particularly in black communities medical care is virtually nonexistent. President Eisenhower understood that a nation can weaken itself beyond repair in the misguided pursuit of strength:

> No matter how much we spend for arms, there is no safety in arms alone. Our security is the total product of our economic, intellectual, moral, and military strengths.
>
> Let me elaborate on this great truth. It happens that defense is a field in which I have had varied experience over a lifetime, and if I have learned anything, it is that there is no way in which a country can satisfy the craving for absolute security—but it easily can bankrupt itself, morally and economically, in attempting to reach that illusory goal through arms alone. The Military Establishment, not productive of itself, necessarily must feed on the energy, productivity, and brainpower of the country, and if it takes too much, our total strength declines.

A nation that puts its trust in military power to solve its problems weakens itself in many ways. The most disastrous consequence of building the power of military institutions at the expense of the rest of the society is to turn that society into a Warrior State ready to sacrifice its most precious values, including freedom itself, to defense against foreign enemies real and imagined. Although the militarization of American life is far advanced, we are not at that point. But Japanese history of the 1930's should warn us that when a nation turns over control of its destiny to the military, it has already been conquered.

However, there are more subtle ways to lose freedom through the uncontrolled growth of military institutions. A nation that avoids confronting its own domestic problems must expect mounting civil disorder. There are only two responses. A society that has decided that it cannot afford to deal with the causes of unrest must suppress unrest. If it will not seek a security based on the consent of the governed, then it must seek the security of the detention camp and the graveyard. Eavesdropping is already sanctioned in the name of national security. The Department of Defense is already in the business of gathering intelligence about student protest movements and of coordinating efforts to control city riots. The revelations that the Central Intelligence Agency once penetrated domestic student movements, universities, church groups, and foundations demonstrate how easy it is for military and para-military institutions to extend their power in an atmosphere of fear and secrecy.

Finally, a nation that invests most of its public money in instruments of violence gives its citizens a powerful example in how to solve problems. Law-and-order speeches make little impact when those in authority seek order primarily through force. The people lose faith in peaceful change when governments resort to violence.

This book is a taxpayer's guide to national security. The first chapter examines the military budget to see what the taxpayer is getting for his money. Some of the basic assumptions behind military spending are explored to enable the citizen to decide for himself whether the nation is spending too much or too little on the military. It should come as no surprise to anyone who has read this introduction that I have made up my own mind on the question. However, I hope to share with the reader the thinking that convinced me when I was working in the Department of State and later, briefly, as a consultant to the Department of Defense that our defense policy is dangerously irrational. The second chapter is an explanation of the role of the military-industrial complex in setting national priorities. It describes how the day-to-day operations of a set of institutions which we have allowed to develop undermines America's capacity to solve its real problems. The third chapter offers a strategy for shifting from an Economy of Death to an Economy of Life. It describes a program of national conversion and includes specific actions that citizens can take to bring about such a conversion. It is offered not as a blueprint but as a stimulus to constructive action.

I

The Trillion-Dollar
Misunderstanding:
A Look at the Defense
Budget

How much defense spending is enough?

The question cannot be answered by consulting computers. The answers of the military professional and the defense contractor, each of whom has a vested interest in big defense budgets, must of course be carefully considered. But they should not be uncritically accepted. There is no handy formula by which the nation can allocate a fixed percentage of Gross National Product to the military and assure either an adequate defense or an adequate society.

Tradition has it that defense spending is not excessive as long as it stays below 10 percent of the Gross National Product. This is a reassuring but meaningless yardstick. The enterprising bureaucrat who decided over twenty years ago that the world's richest society should tithe to support its war machine must be surprised to find that his formula is still being applied as the Gross National Product approaches a trillion dollars a year. Today economists are beginning to understand that Gross National Product, defined as the sum of goods and services generated by an economy in a year, is not the same as wealth. Indeed, it includes items that, as Adolph Berle points out, cancel each other out, such as dollars spent on cancer research and dollars spent on cigarette advertising. The Economy of Death itself, which has provided so much fuel for America's postwar boom, produces relatively little real wealth itself, since its product satisfies no social needs.

The idea that more should be spent for defense as the economy expands has become an essential tenet of the new state religion of national security. One who lives by this faith is immune to the shock of astronomical numbers. Former Secretary of Defense Clark Clifford could scarcely keep from congratulating himself for keeping his "austere" and "modest" defense budget down to $81 billion. And austere it was compared with the $30 billion more demanded by the Joint Chiefs of Staff to fund three "urgent" new weapons systems and ten others that were merely "critical." The annual request of the Joint Chiefs of Staff for new weapons is called the "wish list," a designation that nicely captures the Santa Claus spirit that permeates the defense procurement business. The general or admiral who sulks because his pet airplane or missile is not under the tree at budget time poses a problem for his civilian superiors. Sometimes, as Robert McNamara found out, he sulks in public or before Congressional committees.

Because the pursuit of national security through the arms race is a matter of faith rather than logic, arms spending is at present impossible to control. Former Director of the Bureau of the Budget Charles Schultze points out that the $30 billion annual savings that would result from an end of the Vietnam war cannot automatically be rescued for the civilian economy. In fact, most of the projected "Vietnam dividend" is already committed to new military programs. Unless some crucial assumptions behind present defense policy are explicitly rejected, the Pentagon's escalator will soon take the American public on a ride toward a $200 billion annual budget. Senator Stuart Symington, veteran member of the Armed Services Committee and former Secretary of the Air Force, points out that even a serious try at building a "thick" anti-ballistic missile system would cost about $400 billion. In five years a $200 billion defense budget is likely to sound as austere as $80 billion does today. As for those who wish to return to the $50 billion Eisenhower budget, the staggering size of which prompted the retiring President to utter his famous warning against the "military-industrial complex," such would-be budget-cutters are viewed in the Pentagon as proponents of unilateral disarmament.

There is no way to fix a rational limit to defense spending other than by the application of old-fashioned political judgment and moral insight. Unless the American people begin to ask and keep asking what real security they are buying, there is no hope of stopping the mindless expansion of the war machine. For a generation these questions have not been asked. Until the Vietnam war and the fight over the anti-ballistic missile, no significant interest group in or out of Congress challenged the basic assumptions behind any project of the military, no matter how massive the investment. Every new weapons system has been presented to the public doubly wrapped: an inside wrapping of baffling technical detail, and on the outside, the flag. When the Joint Chiefs of Staff proclaim a new military "requirement" based, as they like to point out, on their 178 years of collective military experience, the taxpayer is expected to say thank you for being taken care of so handsomely.

Those in charge of the Pentagon have treated the defense budget as a mystery. They have protected their mystery as well as themselves with a network of secrecy and classification laws. Long immune from political debate, they periodically terrorize the public with oracular predictions of doom. In 1949 they hinted that 1952 was a "year of maximum danger." Then it was 1954. Then the military managers looked into the void and saw a bomber gap, and then a missile gap, and then an anti-missile gap, none of which turned out to be real.

If a taxpayer is interested in understanding where 70 cents of his tax dollar goes and whether it is being spent wisely or not, there are two kinds of questions to ask. The first is whether the assumptions on which present levels of defense spending rest are correct. The second set of questions concerns the process by which the decisions are made. In this chapter we shall examine the fundamental assumptions of defense spending. In the next we shall take a look at the military-industrial complex to find out how the defense budget is made.

How Many Nuclear Weapons Do We Need?

The principal premise of the defense budget is that more nuclear weapons mean more national security. The United States nuclear arsenal now contains 1000 Minuteman missiles located within the United States in concrete silos, each at present equipped with a one-megaton warhead (fifty times the destructive power of the atomic bomb that incinerated Hiroshima). In addition, the nation has 54 Titan-2 missiles, each capable of equal or greater destruction. Our 41 nuclear-powered submarines can fire 656 more nuclear warheads from submerged positions that cannot be detected by the enemy. Our 646 heavy bombers (B-52's) can carry two twenty-megaton weapons each. A portion of this fleet is on airborne alert at all times and thus is not vulnerable to an enemy surprise attack.

The Soviet Union has 150 intercontinental bombers and about 1000 ICBM's. According to former Secretary of Defense Clark Clifford, the United States, as of September 1968, had about 4200 nuclear warheads aimed at the U.S.S.R., and the Soviets had 1200 that could land on the United States. A year later the score stood at 4500 to 1700. The "superiority" of the U.S. in nuclear striking power is actually greater than indicated, since American naval forces and tactical aircraft operating in the Mediterranean and the Pacific can also threaten Soviet territory with nuclear weapons. The Soviets, except for the brief Cuban adventure in 1962, keep their nuclear weapons on their own territory.

The United States spends about $10 billion a year to maintain and increase the number of nuclear weapons it can land on the Soviet Union beyond the 4500 already on the firing line. Emulating Detroit's ideal car-purchaser, the Pentagon is very close to embracing the annual trade-in. Minuteman I, which was installed in the 1963–65 period and was partially replaced in 1967 by Minuteman II, is now being totally replaced by Minuteman III. Polaris is being replaced by Poseidon. Both new systems are to be equipped with MIRV (Multiple Independently Targeted Reentry Vehicles). This means that the new systems can hurl several nuclear weapons at once in several different pre-selected directions. If these programs are completed, the U.S. will have more than 11,000 nuclear weapons targeted on the Soviet Union. Each submarine commander will have the power to wipe out 160 cities.

The greatest danger of making a religion of national security is that it discourages the application of either reason or experience to human affairs. For a generation it has been enough to justify additions to America's monumental nuclear arsenal by pointing out that the Soviets have nuclear weapons and are building more. Secretary of Defense Melvin Laird recited the catechism perfectly in a speech made only a few months after taking office: "If the Soviet Union is developing a capability that could endanger this nation, we must be prepared to counteract it." The United States, he continued, should err on the "safe side" by "overestimating" rather than "underestimating" the threat. The speech, which was part of an all-out campaign by the Nixon Administration for the ABM, sounded unusually alarmist, but its reasoning was quite orthodox. In his statement accompanying the Johnson Administration's final defense budget, Secretary of Defense Clifford called for the deployment of MIRV, the multiple-warhead system, and other major additions to the nuclear forces, as well as billions more for research and development of still more nuclear systems. All this new hardware was specifically designed to meet "greater than expected threats"—i.e., to "counter" weapons the Soviets *might* build which, according to the best information from our intelligence and communications agencies (a service for which the taxpayer paid a little more than $6 billion in 1969), they *won't* build. "We cannot gamble on estimates of Soviet intentions," Secretary Laird has said, echoing each of his predecessors.

These notions are, of course, a recipe for an unlimited arms race. Since there is almost no weapons system remotely imaginable which the Soviet Union could not build if it devoted enough time, energy, and resources, the only limit to U.S. military spending is the Pentagon's imagination. In the real world, individuals who spend most of their money to arm themselves against threats that exist only in their minds are called paranoid. In the world of national security, the system itself is paranoid. The fear of being "behind" is so great that those who sacrifice scarce resources to counter fictitious bomber gaps and missile gaps acquire reputations as prudent managers. In its haste to counteract its own self-made nightmares, the Pentagon ignores its own experience. According to Senate Majority Leader Mike Mansfield, the U.S. has spent $23 billion on missile systems that either were never deployed or were abandoned. In a paranoid system, waste is a way of life.

But, as has been pointed out, even paranoids have enemies. There is no doubt that the Soviet Union has the capability of destroying United States society and at least one half of our population. That is the uncomfortable reality with which this country has been living since the mid-1950's. The issue is not whether the Soviet Union is or is not a dangerous power to coexist with on the planet. Unquestionably the Soviet Union is dangerous. Any power that is in a nuclear arms race with us is a threat. The real issue is whether building more nuclear weapons increases or decreases the threat to national security.

In strictly military terms, there has been a loss of security. Twenty-five years ago the territory of the United States was invulnerable. Today, as former Secretary of Defense Clifford candidly reminds us, neither the U.S. nor the Soviet Union "could expect to emerge from an all-out nuclear exchange without very great damage—regardless of which side had the most weapons or which side struck first." With each passing year of the arms race the Soviet Union has acquired the capability of killing more Americans.

Years ago the military lost the ability to defend the people of the United States.

During the course of the postwar arms race the world has become in many ways a more dangerous place for Americans to live. Besides the U.S. and the U.S.S.R., three other nations—France, Britain, and China—are stockpiling nuclear weapons. Ten more nations are on the threshold of becoming nuclear powers, among them nations like Israel and Egypt with powerful internal pressures to use such weapons for their very survival. Military laboratories around the world are moving briskly ahead with "contingency plans" for militarizing the ocean floor, fighting nuclear wars in space, raining clouds of anthrax germs on populations, and creating artificial earthquakes. Most of these laboratories are in the United States. But the fact that the U.S. is the chief contestant in these arms races of the near future does not mean that the American people are safe.

Many of these new technological developments will be cheaper and more available than the old. Small countries will acquire them, and so will private groups like the Mafia. Within ten years it should be quite possible to purchase a small nuclear weapon on the private market. When any one of fifty submarine commanders has the physical power to destroy 160 cities, as will soon be the case in the U.S. Navy, nuclear proliferation will be an accomplished fact. Whether every commander recruited over the next generation will resist the temptation to solve the Russian threat or the Chinese threat on his own or to push the button in some future *Pueblo* incident, no psychiatrist can tell us in advance. It will not be pleasant waiting to find out.

Yet the assertion that more nuclear weapons mean more security sounds logical, for throughout history it has been better for a nation facing an enemy to have more weapons rather than less. But that logic has been swept away by two stubborn facts that cannot be eradicated by all the money the Pentagon will ever spend. The first is that the only thing you can do with a nuclear weapon that will promote security is to hold it as a deterrent to others who might drop one on you. The second fact is that any attempt to do more with it, such as blackmailing an adversary, will lead either to a nuclear war or to an endless arms race—in short, less security, not more.

Before the nuclear age a nation could calculate its killing power, measure it against that of its enemy, and make a rational judgment whether to go to war. To conquer Alsace-Lorraine, for example, was worth so many thousand German soldiers. But such a judgment is absurd in the world of the atomic bomb because there is no rational political objective worth the destruction of your own society. Unless it is being run by certifiable madmen, a state will not commit its whole population to suicide—or even a substantial fraction of it—for gold, honor, land, ideological victories, or anything else. The age-old game of international "chicken," in which one nation defined itself in terms of its victories over another, is an historical relic. The U.S.S.R. has three or more one-megaton warheads in missiles aimed at every major American city. As the last three U.S. Presidents have candidly stated, there is no defense against a full-scale Soviet attack.

Deterrence: How to Scare Other Countries

What is necessary for effective deterrence? According to the Secretary of Defense, deterrence depends upon the power to bring "assured destruction" to the enemy. The Soviets must know that if they launch a surprise attack against the United States, they will be destroyed. They must be scared out of doing what they have the physical power to do.

How many weapons does it take to do this? The Soviets have five major metropolitan areas and 145 other cities with a population in excess of 100,000. Beyond these there are mostly villages. One hundred nuclear weapons landing on the Soviet Union would, for practical purposes, destroy the society. According to figures released by the Department of Defense, 37 million people, or 15 percent of the population, would die and 59 percent of the industrial capacity would be destroyed. Assuming that larger numbers of U.S. nuclear warheads land on the Soviet Union in a retaliatory attack, the arithmetic of death assumes these proportions:

Number of Warheads	Soviet Fatalities		Industrial Destruction
	Millions	Percent	Percent of Capacity
200	52	21	72
400	74	30	76
800	96	39	77
1200	109	44	77
1600	116	47	77

It is highly improbable that any Soviet leader would deliberately court an attack if he thought there was even a substantial risk of 100 nuclear weapons falling on the Soviet Union. Indeed, if we accept that the Soviet leaders are political men whose primary job is trying to figure out how to govern Russia rather than to destroy the U.S., they are likely to be deterred by far less. It is inconceivable that the present Soviet leadership or any sane successors would deliberately sacrifice even one city for the mere purpose of challenging or humiliating the United States. Unlike a former U.S. Secretary of the Navy and a dozen or more Air Force generals, the Soviets have never publicly threatened preventive war. As the invasion of Czechoslovakia makes clear, many Soviet leaders are ruthless men ready to use force for specific and limited political ends. But they know their own interests. These interests, as President Nixon pointed out in a televised press conference, have traditionally prompted them to emphasize defensive rather than offensive weapons in their military forces. Until recently, in fact, they have appeared willing to live indefinitely with a much smaller nuclear arsenal than the U.S. Above all, they are conservative in courting risks to the Russian homeland. The memory of the 25 million Soviet citizens who died in World War II is still fresh in the Soviet Union today.

However, in order to be sure that 100 weapons get through the powerful Soviet air defenses and actually kill people by the millions, let us assume that the "assured destruction" force should be 300 or 400. Then double that figure as "insurance" against misfires. Such a force would already represent incredible "overkill"— i.e., far more killing power than could conceivably be needed for deterrence. The prospect of even a fraction of that force would impress any rational leader, and nothing is enough to impress an irrational one. The U.S. nuclear arsenal is now more than five times larger than this. Additions to that force already programmed will make it ten times larger and more. There is no reason for maintaining a nuclear force of this size if the purpose is deterrence only. A country that feels secure only if it has the power to kill its enemies 100 times over has problems that additional missiles will not cure.

But deterrence is not the only purpose of the nuclear force. The U.S. has tried and is still trying to use its nuclear weapons for other purposes. This is the heart of the matter.

Since the early days of the Kennedy Administration the United States has cultivated the illusion that nuclear weapons can be used in a "counter-force strategy" to support what nuclear strategists call "a credible first-strike posture." In plain English, this means that the U.S. should build a missile force big enough to wipe out Soviet missiles on the ground. Such a strategy, of course, requires the U.S. to attack *first*. The prospect of building such a force delighted an aspiring Clausewitz like Herman Kahn, and, for a while, Robert McNamara too, because it purported to give a rational political purpose to nuclear warfare. With a credible-first-strike strategy the U.S. could threaten the Soviets with a nuclear attack if they did something the U.S. didn't like and make them believe that the threat would be carried out. The Soviets would lose most of their retaliatory missiles in a multi-megaton Pearl Harbor and would thus be able to inflict no more than "acceptable" damage on the United States.

In July 1961, at the height of the Berlin crisis, the Kennedy Administration seriously proposed a massive fallout-shelter program that, nuclear strategists argued, would make "credible" the "will" of the United States to risk nuclear war to keep the Russians out of Berlin. *Life* Magazine assured the public that 99 percent of the population could be saved, and that a nuclear attack would be "just another war." Secretary McNamara then conducted a series of studies which convinced him that a "credible first-strike posture," even with shelters, could not save the U.S. from utter catastrophe. In 1964 he told a Congressional committee that the Soviets' decision to place their missiles on submarines and in concrete-protected underground silos now made it

increasingly difficult, regardless of the form of the attack, to destroy a sufficiently large proportion of the Soviets' strategic nuclear forces to preclude major damage to the United States, regardless of how large or what kind of strategic forces we build. Even if we were able to double or triple our forces we would not be able to destroy quickly all or almost all of the hardened ICBM sites. And even if we could do that, we know no way to destroy the enemy's missile-launching submarines at the same time.

No critic has come up with any better demonstration of the futility of stockpiling further nuclear weapons. McNamara's logic was backed up by estimates of the damage the United States would suffer under various contingencies. Assuming a massive Soviet build-up of offensive missiles including multiple warheads and a big U.S. ABM program, the Department of Defense reports that U.S. fatalities "rounded to the nearest 500,000 people" would be between 25 million and 40 million. That would be the happiest possible outcome. Under other more likely contingencies the casualties would range from 80 million to 110 million. Thus the "credible first strike," otherwise known as "nuclear blackmail" when talking about possible Soviet use of the same strategy, is not only genocidal, but suicidal. The attempt to turn a "superiority" in numbers of weapons and tons of killing power into political victories was doomed once the Soviets could destroy 25 million or more Americans.

But the Pentagon strategists were undaunted. They refused to believe that nuclear weapons were effective only for deterrence. If that were true, there was no reason to maintain the present nuclear force at a cost of $3 billion to $4 billion a year, much less to triple it. But although the generals in charge of the nuclear forces did not know how to win nuclear wars, they did know how to win arguments. They developed a series of sophisticated rationales for a never ending increase in the nuclear arsenal. Sensitive to charges of "overkill," the proponents of the nuclear build-up argued that the new weapons were not for killing more people, but only for making the enemy *think* that the U.S. was going to kill enough. Under such a theory, additions to the strategic forces designed to buy "assured destruction" should really be part of the education budget. Assuming the expenditure is for the enlightenment of the members of the Soviet Praesidium, the per-pupil cost is about $1 billion per year.

Damage Limitation: How to Make Sure the Arms Race Doesn't Stop

Another rationale for an unending build-up is "damage limitation." According to this theory, the U.S., in the event of a Soviet first strike, might be able to knock out some of the Soviet missiles left on the ground before they were fired and so reduce American fatalities. In a burst of candor Secretary McNamara undermined the basic premise of this strategy too. "I personally believe that it is extremely unlikely that in the event of a nuclear strike by the Soviet Union against this country they would do anything other than to strike all of our nation with all their power in the very first strike." On May 2, 1968, Dr. John S. Foster, Jr., Director of Defense Research and Engineering, reiterated the same point. The Soviets would "use all the forces they have" in any first strike on the U.S. So the extra missiles the U.S. needed for "damage limitation" would strike empty holes.

Nevertheless, it is possible to imagine some circumstances, all of them fairly implausible, in which a "damage-limiting" force might somewhat reduce U.S. fatalities in a nuclear war. A deranged Soviet officer fires off a single missile, or a Soviet submarine commander pushes the wrong button and blows up Seattle. In such a situation, according to the Pentagon scenario-writers, if the U.S. has missiles aimed at most of the Soviet missiles and fires before the Soviets fire theirs, U.S. casualties might be held to between 30 million and 40 million persons. Perhaps as many as 50 million might be saved if the Soviets were willing to provide a war that conformed perfectly to the Pentagon's rules.

The argument for building ever more missiles to reduce damage is appealing. Who can be against saving millions of lives? The superfluous missiles are defended as "insurance." But no analogy is more misleading. Buying an insurance policy has no effect on what happens to the individual. If you buy too much insurance, you make a bad investment, but you don't hasten your own death. Building new weapons systems, on the other hand, is a political act that creates a war climate and insures a stepped-up arms race.

How does the arms race work?

The experience of the last twenty-five years makes

clear that an attempt to stockpile nuclear hardware beyond what is conceivably needed to destroy the adversary's cities sets off a frantic escalation of the arms race on both sides.

Imagine the following conversation in the Kremlin. A Soviet general with a secret intelligence report on U.S. war plans gleaned from a careful reading of *The New York Times* demands an urgent audience with his Chairman. He then informs him that the U.S. is proceeding to increase its nuclear-missile stockpile from 4500 to 11,000 by building MIRV, the hydra-headed missile that can hit several pre-selected targets at once, and at the same time is also building ABM.

"Why are they doing a thing like that?" asks the Chairman.

"Don't have a clue," replies the general. "Or, rather, we've got too many clues. The Americans give a new reason for building these forces every time someone shoots down the old reasons."

"I don't like it. Do you think they are trying for a surprise attack?"

"They'd be crazy to try even with twenty thousand nuclear warheads. They could never knock out all our retaliatory missiles. But maybe they don't know that. They say they do, but you can't trust them. We must assume that MIRV might destroy nearly all our present missile force. We'd better err on the safe side and build both MIRV *and* ABM. Nothing less will convince the Americans that getting into a nuclear war with us is dangerous."

Meanwhile, back in the White House, Secretary Laird lays a sheaf of top-secret photographs before President Nixon: "Looking over the development in the current deployment of the SS-9 [the late-model Soviet intercontinental missile] leads me to the conclusion that with their big warhead and the testing that is going forward in the Soviet Union, this weapon can only be aimed at destroying our retaliatory force."

"This looks like a security gap to me," says Nixon. "How many do they have?"

"They'll have maybe five hundred by 1976. It looks like they're trying for a first strike, and make no mistake about that. We need to increase our yearly spending on nuclear forces by at least ten billion."

The only difference between the two conversations is that while the Kremlin discussion is imagined, the White House discussion is reconstructed from actual public statements. Because each side is uncertain about the intentions of the other and the consequences of a "degraded deterrent" are so disastrous, both overdesign and overbuild their forces. When one side builds too much, the other assumes that it too must prepare for the worst. When both sides arm themselves against their own strategists' most pessimistic fantasies, the arms race spirals. Thus the attempt to gain absolute safety in a nuclear war by building more weapons makes nuclear war more likely. When both sides are in crash programs, there is a growing risk that one or the other will panic and try a "pre-emptive attack," however irrational. One of the many specious arguments for building more missiles and particularly the ABM is to protect against a small attack launched by an "irrational man." It is hard to think of a more unstable atmosphere for an unstable man than the continuing arms race.

MIRV is peculiarly dangerous because it increases uncertainty on both sides. Neither the Pentagon nor the Kremlin knows how many separate warheads are concealed in its adversary's missiles. Each plans for the worst and thus makes a reality of the other's nightmare. The losers are the American and Soviet peoples. Both are taxed in new and higher amounts to support a system in which the only sure result is that in the event of nuclear war more people will be killed. Since the actual number of warheads concealed in a missile nose-cone is unknown, the introduction of MIRV greatly complicates the problem of inspection and makes arms control almost impossible. As Clark Clifford put it in his final statement as Secretary of Defense:

> We stand on the eve of a new round in the armaments race with the Soviet Union, a race which will contribute nothing to the real security of either side while increasing substantially the already great defense burdens of both.

With this regrettable development noted, the Secretary went on to list his "strategic requirements." In addition to MIRV for the land-based Minuteman, the taxpayer needs Poseidon, which is a MIRVed version of Polaris, to shoot from submarines; SRAM (Short-Range Attack Missile) to launch from bombers to overcome Soviet air defense and to protect the bomber from any suggestion of obsolescence; SCAD (Subsonic Cruise Armed Decoy), another aspect of the bomber-rejuvenation program; ULMS, a new sea-based missile system for use when it is trade-in time for Poseidon; and the marvelously resilient advanced manned bomber, which appears to be impervious to rejection. This hardware list, which is being expanded by the Nixon Administration, costs about $10 billion a year, and within the next five years the annual cost is likely to be almost twice that amount. The total cost of these programs could come to more than $100 billion in all. That is a sizable investment for something that "will contribute nothing to the real security" of the American people.

How to Stop Worrying and Love the Bomb

Why is it that America's leaders are willing to say that more nuclear weapons mean *less* security but are unwilling to act on their own logic? Speaking at the United Nations in 1961, President Kennedy eloquently stated the ultimate reality of the nuclear age: "Mankind must put an end to nuclear weapons or nuclear weapons will put an end to mankind." In the same year he proceeded to raise the defense budget by almost $20 billion.

There are two principal rationalizations for the institutionalized suicide that we call the arms race. The first is the "it's a bad bad world" syndrome. Man is so steeped in sin and the planet is such a dangerous place to live that there is "no alternative" but to keep piling up bombs, even though a minute's cool reflection makes clear that this leads to even worse "alternatives." The hard-nosed political realist dismisses the fears of the leading scientists and ordinary citizens alike about the nuclear arms race and counts on the power of positive thinking to get humanity through the rest of the century. He sees no connection between his behavior and the behavior of others. If only the Russians weren't so aggressive and the Chinese weren't crazy. It would be nice if the world were a place where one didn't have to keep arming. But it isn't. Until it gets better, we are going to keep on making it worse.

The United States has the greatest aggregate of economic and military power in the world. For this reason the United States military policy often has a decisive influence on others. It is now clear that the U.S. has taught the Soviet Union most of what it knows about what to do with nuclear weapons. Soviet strategists have consistently imitated their American counterparts, usually two or three years later. We have taught them, contrary to their own inclinations, that it is better to have more weapons than less, and that every technical improvement must be put to military use. The result is that the Soviet Union now devotes far more of its resources to military purposes than in former years.

According to the Joint Economic Committee of the Congress, between 1950 and 1962, following the first major U.S. postwar rearmament, the Soviet military quadrupled their expenditures. In the years 1963 to 1969 the Soviet Union has more than quadrupled its missile force. Again this *followed* the U.S. build-up of the Minuteman and Polaris missiles of the early 1960's. The size and character of the U.S. investment in nuclear weapons directly influences the Soviets' response. Since their resources are much more limited, the effort to catch up with the U.S. requires even more militarization of the economy than in the U.S. It means giving added power and money to an increasingly powerful class of specialists in violence who live off the American threat just as the Pentagon lives off the Soviet and Chinese threats. In neither country do the military have any real incentive to see these threats diminish. Their ascendancy portends more aggressive policies and a greater likelihood of war. It is hardly a national-security triumph that the Soviets are becoming "more like us" in stockpiling nuclear weapons and in the increasingly bold use of arms diplomacy in such places as the Middle East and Cuba. Through its arms policy the U.S.

has helped build an enemy.

The proponents of a permanently expanding nuclear arsenal have a second rationalization which they use with equal success to protect the military establishment from the real world. Winston Churchill's slogan "We arm to parley" is their official credo. Once the United States builds up its military forces to some unspecified level, the Soviet Union will become readier to negotiate. We must never enter negotiations from a position of "weakness." This rationalization for the permanent arms build-up ignores twenty-five years of history. In the first place, the United States, as we have seen, has always had an enormous superiority over the Soviet Union in numbers of nuclear weapons and still has today. If we put ourselves for a minute in the shoes of the Soviet leaders, it becomes instantly clear why such "strength" cannot be translated into arms-control agreements. No nation wants to stop when it is behind, particularly if the adversary appears to be trying to build up a force capable of launching a surprise attack. Indeed, the U.S. build-ups have stimulated the Soviets to aim more missiles at the U.S. than they wanted to. In the 1950's, when Nikita Khrushchev was Premier, the Soviet Union preferred to base its deterrent on threatening speeches rather than hardware. The Soviet leader would periodically claim that he was producing missiles "like sausages" when actual production was about 2 percent of what we assumed the Soviets could build. He boasted that Soviet forces could "hit a fly in outer space" with an anti-missile that didn't exist. With an assist from U.S. politicians who exploited the issue to run for office, Khrushchev created a totally false picture of a missile gap. In the short run it was very successful. Khrushchev understood that the object of the nuclear arms race was not more bang for the buck, as former

Secretary of Defense Charles Wilson put it, but more terror for the ruble. He had better uses for his money than to put it into missiles. In the long run, however, with the development of better U.S. spying techniques, the policy of deterrence through boast failed. The relative weakness of the Soviets was exposed. The Soviets went into a crash missile program to catch up. Now that they may be nearing the completion of present programs, their interest in arms control has revived. But the U.S. now plans to triple its nuclear arsenal with MIRV. And so on into the next round.

The plain truth is that these rationalizations for the arms race are not the real explanation. Totally conflicting signals from the enemy produce the same result. If, as the missile-gap enthusiasts argued in 1960, the Soviet Union is ahead, then we must go into a crash program to catch up. If, as the Kennedy Administration found within a month of taking office, we are ahead, then we must preserve our "superiority." If, as the events of the past three years have revealed, the Soviets are unwilling to accept permanent inferiority and are running a crash program of their own, then we must redouble our efforts. In each case the analysis of what is happening in the Soviet Union is different. In each case the prescription is the same: more. As we shall see in the next chapter, what the U.S. spends on nuclear weapons is determined far less by Soviet behavior than by America's own inner drives.

The General Purpose Forces:
Do We Need to Prepare for Three Wars at Once?

Although the nuclear arsenal accounts for most of America's killing power, only a small fraction of the taxpayers' money goes to buy nuclear weapons. Most of it goes to the so-called "General Purpose Forces," which include all the men and equipment in the military establishment except those with the mission of assuring the atomic destruction of Soviet and Chinese societies. These forces, which now cost more than $50 billion a year, have been steadily increasing since 1961.

There are more men under arms in the U.S. today than in either the Soviet Union or China. The U.S. Army has 18 divisions and 5 brigades. The nation maintains over 33,000 aircraft located in bases on every continent. The Navy consists of 15 carriers, each equipped with a vast array of nuclear weapons; more than 100 attack submarines; 240 escort destroyers; and thousands of smaller craft.

The United States General Purpose Forces constitute a greater aggregate of so-called "conventional" military power than that possessed by any other nation. The American forces in Europe rank as the third-largest military establishment on earth. The National Security Council has recently considered among several alternative "options" the possibility of increasing the budget for General Purpose Forces from $50 billion to $85 billion a year. According to the Secretary of Defense, the $80 billion in the 1970 budget will give us "balanced forces." It is easy to see why. The money is divided about equally among the Army, Navy, and Air Force.

About 50 cents of each tax dollar goes to the General Purpose Forces. What is the taxpayer getting for this investment? What are the general purposes of the General Purpose Forces? Do they make sense? Unless the Congress and the public are prepared to ask these questions and to demand answers, the analysis of defense spending will not penetrate beyond the accidental waste of bad management into the inherent waste of the present national-security system itself.

It is important to understand that the General Purpose Forces are useless for defending the continental United States from nuclear attack. Since there is no power on earth with enough ships, planes, men, or nerve to mount any other kind of attack on this country, the "security" for which $50 billion a year is invested must mean something other than national defense.

Why does the U.S. need to maintain a permanent military force of such size?

The defense budget is explicitly based on the assumption that the U.S. must be prepared to fight three wars simultaneously: a major war in Europe, a substantial war in Asia, and a somewhat smaller war elsewhere, either in Latin America or in the Middle East. Based on these assumptions, the defense budget is too low. It is not large enough to win the war in Vietnam, much less carry on two more at the same time. Planning for three wars at a time justifies unlimited military spending.

But why does the U.S. need to plan on fighting three wars at once?

There are two types of wars for which the U.S. is preparing: conventional war, where one nation attacks another, and counter-insurgency war, where the U.S., as in Vietnam, aids a government in putting down a revolutionary challenge. The U.S. has made commitments to forty-two countries around the globe to defend them against outside aggression. Several of these commitments have been made by the President without obtaining the approval of the Senate as is provided for in the Constitution. Except in one or two cases, the interest of the American people in these commitments has never been publicly debated. They all were undertaken years ago in radically different circumstances. Except in Korea, with which we had no treaty, the U.S. has not actually fought a conventional war in the entire period since World War II.

These commitments are centered in Asia and Europe. According to the Department of Defense's own assessment, U.S. forces are not needed to defend India, Taiwan, or Korea. Indian forces of 1.1 million men "should be able to defend their country against Chinese aggression." In Taiwan, Nationalist Chinese land forces (372,000) are "fully adequate" to defeat a Chinese amphibious assault. As for Korea, Republic of Korea (ROK) land forces "provide a strong deterrent against even a maximum Chinese/North Korean attack."

The Vietnam experience suggests that the U.S. does not have the power to defend a country like Laos, Cambodia, or Thailand from a full-scale attack by North Vietnam or China. Nor is such an attack likely. Military planning that fosters the illusion that the U.S. can enhance either its own security or anyone else's by fighting a major land war in Asia is not just expensive; it threatens to involve the American people in another catastrophic conflict 10,000 miles away without political or moral purpose. The idea of maintaining a permanent "U.S. military presence" in Southeast Asia after the Vietnam war is over should be explicitly rejected, as Generals Shoup, Ridgway, and others have strongly advised. If that were done, the nine U.S. divisions presently in Asia could be demobilized. A leading Pentagon consultant has estimated that if we would plan for only two wars instead of three, we could save $5 billion a year.

The United States has maintained five combat divisions and their dependents in Europe for almost twenty years. The cost per serviceman in Europe as elsewhere is about $10,000 a year. Each combat soldier must be backed up by seven or eight more men. Thus it costs between $80,000 and $90,000 a year to put one man in the line. The NATO countries, without even counting U.S. troops in Vietnam, have over 2 million more men under arms than the Soviet Union and her allies. NATO has more troops on the line in Europe than the Warsaw Pact nations, and about 2000 more aircraft. The 310,000 American troops and the 200,000 dependents are far more than what is needed to symbolize the U.S. commitment to oppose a massive Soviet invasion, and not nearly enough in the exceedingly unlikely event that the Soviets should try one.

As Senator Stuart Symington of the Senate Foreign Relations and Armed Services Committee says, "Surely 50,000 American troops would be sufficient to make sure that no Soviet probe could succeed in Berlin or elsewhere in Europe without a direct confrontation with the United States." Robert Benson, formerly in the office of the Comptroller of the Pentagon, has pointed out that even reducing the troops in Europe by a modest 185,000 would save $1.5 billion a year. To implement Senator Symington's proposals, which General Eisenhower also once endorsed, would save substantially more. Tactical nuclear weapons in Europe, the use of which would totally destroy the territory to be defended, alone cost well over $3 billion a year.

Counter-Insurgency: The World Is an Expensive Beat

By far the greatest part of the General Purpose Forces is maintained not to defend our allies from outside attack but to defend their governments from their own people. The U.S. intervention in Vietnam is a prime example of the attempt to use outside military power to put down a local revolutionary movement. As any television viewer can see with his own eyes, the attempt has failed, politically and morally. More than $25 billion a year is budgeted to carry on the war in the name of U.S. national security. After a counter-insurgency campaign of fifteen years and a full-scale war of five years, which have cost more than $100 billion and almost 40,000 American lives, the average citizen is in a position to evaluate the effort for himself. "Our forces have won every major battle in which they have been engaged since their commitment in South Vietnam," former Secretary McNamara told the Congress as late as April 1968. Then why has the war not been won?

The clearest lesson of the Vietnam war is that the concentration of military power against even one of the world's weakest countries cannot be translated into political power. The United States can destroy Vietnam with our bombs, but we cannot create a Vietnamese society to our liking. The unsuccessful attempt to coerce Vietnam into political subservience to American interests has only revealed our impotence. The "capabilities" which we buy to carry on self-defeating wars are negative assets.

Yet the taxpayer is asked year after year to spend billions for destruction that serves the interest of no one other than bombardiers in training and manufacturers of napalm. In 1969–70, for example, the U.S. budgeted $5 billion for high explosives to drop on Vietnam. The Pentagon plans to dump about 90,000 tons of ground ammunition and 110,000 tons of air-delivered munitions each month on South Vietnam in a stepped-up crusade to save it. That comes to 2.5 megatons per year, or 125 times the destructive power of the atomic bomb that leveled Hiroshima. In the first three months of 1969 the hail of bombs actually dropped exceeded these projections.

However, as official explanations of the defense budget make clear, Vietnam is not a unique situation, only an unusually difficult one. The U.S. objectives in Vietnam and the strategy for realizing them have been extended to many other places. For more than twenty years the U.S. has carried on a global campaign against revolution and native insurgent movements, conducting a major military campaign or a CIA operation in an underdeveloped country about once every eighteen months. Greece (1948), Iran (1953), Guatemala (1954), Lebanon (1958), Cuba (1961), Congo (1964), British Guiana (1964), Dominican Republic (1965) are a few of the principal examples. The U.S. also took charge of the campaign against Che Guevara in Bolivia and supplied men and equipment to defeat guerrilla movements in Peru, Colombia, and other places in Latin America.

Professor C. E. Black predicts in *The Dynamics of Modernization* that there may be "as many as ten to fifteen revolutions a year for the foreseeable future" in poor countries where the average person earns less than $50 a year. Many of those revolutions may be organized by people calling themselves communists. The U.S. is presently committed to opposing them and maintains enormous military forces for this purpose. If the crusade against revolution continues, U.S. violence abroad will escalate still further. We will find ourselves once again facing the Russians or the Chinese in some future Vietnam where the nuclear giants are led step by step by their clients into a direct confrontation. Once again the U.S. will find itself supporting military dictatorships and corrupt regimes that cannot rule without our guns. Again and again America's war on revolutionary change will require her to make a desert of the primitive societies she claims to rescue. Increasingly, a majority of people on the planet will come to look upon us as Public Enemy Number One. And they will be right.

To act as the Guardian at the Gates, to use President Johnson's term, and to protect the globe from revolutionary change, the U.S. must continue to spend more on its General Purpose Forces, not less. As long as having a military "presence" in Asia, a ring of military bases around the Soviet Union, a five-ocean Navy, and a world-wide counter-insurgency program is considered "vital to the security" of the United States, then there are few economies indeed that can be made. The taxpayer will continue to pay for the support of foreign armies, the training and supply of foreign police, the maintenance of huge bases, and the protection of these far-flung commitments with complex and fantastically expensive intelligence and communications networks.

The whole enormous effort is based on one major premise. When internal revolutionary movements overthrow corrupt, oppressive governments, they somehow threaten the security of the United States. Why should this be so?

A revolutionary regime in a backward country can pose a military threat to the American people only if the U.S. intervenes against it with military force and in the process comes up against the Soviet Union or China. If a country in Asia, Africa, or Latin America "goes communist," the Soviet Union or China does not necessarily benefit. Communism is not a unified world force. The Soviet Union and China are actually fighting each other, and the level of tension within the rest of the communist world—between Russia and Czechoslovakia, to cite an obvious example—is very high. Where revolutionary regimes come to power, as in China, Cuba, and Yugoslavia, they are not puppets of the Soviet Union; revolutionaries are fired by a passion for independence and a desire to be free of all domination by big countries of whatever ideology. Nor are the peasant revolutionaries of Latin America and the Vietcong part of the same army. They will become one only if the United States succeeds in unifying insurgent movements around the world by making itself the common enemy of all. Lyndon Johnson's battle cry "We must fight them in Vietnam now or we will have to fight them in San Francisco tomorrow," evoking a picture of a fleet of sampans sailing under the Golden Gate Bridge, is a psychotic fantasy. Revolutionary regimes, even communist ones, do not make the United States any weaker militarily or the communist powers any stronger.

Nevertheless, the U.S. has two other reasons for playing world policeman. One is a matter of pride and the other a matter of cash. For a generation U.S. leaders have talked themselves and others into the belief that saving the underdeveloped world from communism, by force if necessary, would rescue it from poverty, disease, illiteracy, and atheism. Despite the quantity of rhetoric which celebrates the American Dream of spreading democratic development in the former colonial world, we have brought neither development nor democracy. The number of dictatorial regimes under our support increases year by year, and so does the gap between the rich nations and the poor. The unfortunate truth is that the U.S. has no alternative to offer the poor nations which is any better than revolution, which, for all its brutality, has had some spectacular successes. The rapid modernization of backward Russia and her transformation into the world's second power, the end of massive starvation in China, and the great progress in literacy in Cuba are a few examples of what regimentation and the shake-up of an old corrupt order can do. We may not like it. It is easy to point to terrible evils and injustices. But what about the status quo in many backward countries of the "Free World" where thousands starve, only a few can read, 2 percent of the people take 70 percent of the wealth, and a baby lucky enough to survive its birth can look forward to twenty-five years of a dismal life? It may be that revolution is the only answer to the physical survival of these societies. Since we have no better answer, it is cruel and arrogant to try to impose one. It is also expensive. This nation can bleed itself white trying to prove that suppressing revolution is the answer to world development.

Most of the world believes that the real reasons for U.S. military activity in the Third World are economic. We are trying to preserve our access to raw materials, markets, and profitable investments. In short, we are practicing gunboat diplomacy in modern dress. It is true that where revolutionary regimes come to power, it is usually no longer possible for the United States to carry on trade and investment activities on the same advantageous terms as before the revolution. One of the basic ideas of such radical governments is to stop foreign exploitation of local resources and labor and to try to build a more independent economy. But no revolutionary regime has said that it wants to cut the U.S. off from trade. The issue is the terms. Revolutionary regimes customarily demand higher prices for their raw materials than those they could formerly get in the world market. They insist upon ownership of the precious resources of their land. Yet France was able to work out a successful trade policy with Algeria after she abandoned her attempt to crush the Algerian revolution. It is possible to develop commercial arrangements that are fair to both rich and poor countries.

The American people make up 6 percent of the world's population and consume each year about 60 percent of the earth's consumable resources. This fact underscores the real "security threat" from the underdeveloped world. To live in security in a revolutionary world, the U.S. will have to cope with the unpleasant truth that Americans cannot continue to grow richer while millions starve, and still feel safe. More guns will not help.

The Soviet Communist Party and many prominent American businessmen make the same analysis of the United States. America must preserve its present grip on the economies of the underdeveloped world or else it will have to abandon its high standard of living. There is little doubt that the pattern of consumption in this country will have to change over the next generation, not only to make it possible to distribute resources more equitably around the world but also to prevent world-wide depletion of resources. One thing is clear, however. Whatever benefits the U.S. now derives from preserving economic advantages with military power are purchased at prohibitive cost; the American people cannot afford imperialism, either financially or spiritually, without the nation ceasing to function as a democratic state.

If the use of U.S. military power for counter-insurgency purposes were abandoned, the defense budget could be cut drastically. If we reject the basic premise of present policy—that any regime, no matter how militaristic and repressive, which calls itself anti-communist is our friend and any regime that rejects capitalism and U.S. influence is our enemy—then we would not need such a world-wide show of force.

Here are some specific examples of cutbacks that would not only save money but would improve security. The Pentagon hopes to spend more than $10 billion over the next few years to improve airlift and sealift "capabilities." The C-5A aircraft, a transport for ferrying a battalion of men anywhere on the globe in a matter of hours, and the Fast Deployment Ship, which can transport much larger units, are expressly for the purpose of getting into future Vietnams faster and more efficiently. An almost sure way to repeat the Vietnam disaster is to keep buying equipment for future Vietnams.

Military assistance could also be stopped. These programs have helped a succession of military dictators to come to power and to stay in power. Contrary to the claims made for the programs, they have not promoted stability in Latin America. According to a study of the Senate Foreign Relations Committee, the countries that have received the greatest amounts of military aid have had the greatest number of military coups and violent changes of government. In Greece, a brutal junta used American military aid to stamp out democracy. Military assistance, as its proponents frankly tell the Congress, is to protect poor countries not from external invasion but from their own people. In many parts of the world the U.S. is the sole supplier of arms. Stopping the arms flow would make it necessary for the governments of these countries to figure out how to rule their countries with justice instead of force.

In some countries, chiefly in the Middle East, the U.S. and the Soviet Union are running a race for the title of Number One Merchant of Death. So far the U.S. is far ahead. The U.S. sold military planes to Iran to keep Iran from buying military equipment from the Soviets. Two or three weeks after the shipment was arranged, as Senator Eugene McCarthy has pointed out, "the Iranians went to the Russians for additional military equipment anyway." The Soviet Union has not had much success with its military-assistance program either. In the two countries which have received the greatest military aid, Egypt and Iraq, the local Communist Party is outlawed. A U.S.-Soviet agreement to stop arms shipments to such countries would do much to defuse explosive situations such as that in the Middle East. It would also save a great deal of money for everyone concerned.

The Technology Race

The three most expensive premises of the defense budget, as we have seen, are: (1) the belief that the nation can increase its security by stockpiling more and more nuclear weapons; (2) the U.S. pledge to defend forty-two countries from external attack even if it means fighting three major wars at once; and (3) the world-wide commitment to oppose revolution and guerrilla movements. The fourth principal premise concerns the technology race: the U.S. must always be at the frontier of technology in every field that could possibly have military significance.

Research, development, and testing of new weapons

account for a little more than $8 billion a year at present spending levels. However, the impact on defense spending is far greater than even this considerable figure indicates. When the taxpayer funds a new research project, even at modest levels, he has bought more than a group of energetic scientists and military technicians to invent future weapons. He has bought a lobby. Betting personal careers on a piece of new technology, each cluster of weaponeers measures its worth and prestige in the Pentagon by its ability to increase its appropriation from millions to billions. Every dollar spent this year on a research project is likely to mean our spending five or ten more within a few years. Indeed, the very purpose of military research is to "generate requirements"—i.e., to think of the weapons of tomorrow which in turn will give rise to more weapons for next year. The operating creed in the research-and-development business is simple: "If you can think of a new way to burn, shock, bore, disintegrate, poison, asphyxiate, blow apart, or otherwise persuade potential enemies, research it. If you can research it, develop it. If you can develop it, deploy it." The technology race, to use a term that Pentagon public-relations people like, supplies much of the energy and forward momentum of the arms race.

A massive military research effort is justified on two grounds. First, there is a constant need to scan the horizon for possible dramatic "breakthroughs" that could suddenly alter the military balance. Although such leading weapons experts as former Science Advisor Jerome B. Wiesner have insisted that there is "no technical answer" to the arms race, considerable sums of money are spent each year on the opposite assumption.

The second reason for a big military research budget is modernization. Existing weapons must be continually improved. This year's planes must fly faster than last year's. More lethal germs must be concocted. Tanks and guns must be continually improved. In the world of atomic stalemate and limited war, progress is the Pentagon's most important product. It is a substitute for victory.

The U.S. military establishment has come a long way since the day in the late 1930's when a chemist who wanted to work for the Navy was turned down on the ground that the Department already had one. Today more than half the scientists and engineers in the country work directly or indirectly for the Pentagon. Technological achievement has become an end in itself. The official fascination with the technology of death is accurately captured in this 1964 report of the House Subcommittee on Defense Appropriations:

> The future brings with it . . . , of course, many possibilities. A death ray type of weapon might evolve from "Laser" research. Efforts in chemical and biological warfare might produce incapacitating agents that could render persons helpless for short periods of time without causing large numbers of fatalities. It might be that future military systems will be some type of space-based system —either space platforms from which weapons can be launched or orbiting weapons which themselves can be caused to deorbit and strike targets on command. It may well be that some idea which is currently in the minds of a very few scientists will result in a technological explosion in an entirely new field.

Even though there is apparently some uncertainty as to precise results, the Committee feels that, with wise management, we have the technological base and the level of funding which will enable us to stay ahead in all vital research and development areas.

In his report for 1970 Dr. John S. Foster, Jr., Director of Defense Research and Engineering, states that there are "no areas at present where we know Soviet technology to be significantly ahead of U.S. work." The U.S. is winning the war of the laboratory. Indeed, the Soviets have not been pulling their share of the load. "Soviet military hardware frequently has not reflected the most advanced state-of-the-art in the USSR," Foster notes. The Soviets emphasize "proven technology" and favor "robust, relatively simple equipment." They have stubbornly refused to build an advanced bomber, denying the Air Force a plausible "threat" to justify a host of cherished projects including a "follow-on" bomber of its own. Nevertheless, the Pentagon is now spending over half a billion dollars a year on a series of projects designed to counter the nonexistent late-model Soviet bomber. As Harold Brown, the last Secretary of the Air Force in the Johnson Administration, put it, "The Air Force view is at least as much a view that 'they ought to have one' as it is 'they will have one.' "

Most of the research budget goes to developing and testing new models rather than for basic research into possible breakthroughs. The decision to run the Pentagon in accordance with the ground rules of the automobile economy has resulted in the waste of a great deal of money and the purchase of oversophisticated, ineffective weapons. Detroit's ability to persuade a sizable fraction of car-buyers that it won't do to be seen in last year's model is an important stimulus to the economy; but the annual trade-in has had bad results in the Pentagon.

Take the case of the famous F-111, otherwise known as the TFX. This "all-purpose bomber" for the three services was produced against the advice of both military planners, who wanted a variety of specialized new planes, and civilian planners, who could scarcely think of a mission that the then current model, the F-105, could not do as well. Indeed, as it happened, the technological virtuosity that so excited Secretary McNamara and other enthusiasts of the F-111 turned out to be its undoing. It was too complex to work. Three of the first eight F-111's sent to Vietnam were lost in a matter of weeks; two out of five sent to the Takhli base in Thailand on March 17, 1968, were lost within five days. In all, at least eleven of the aircraft crashed before the Pentagon ordered the plane grounded.

When the Air Force Chief of Staff appeared before the Senate Appropriations Subcommittee, Senator Richard Russell, the Subcommittee's chairman, expressed the hope that the F-111 might fall into Russian hands and be copied. "It would put the Air Force out of business." The plane had become an expensive joke. The original cost of each plane was supposed to be $3 million. Senator Curtis of Kansas, a member of the Aeronautical and Space Sciences Committee, estimates that the planes will end up costing $9 million apiece.

There are numerous other examples where "improvements" in weaponry have rendered them almost useless. The late-model rifle, the M-16, distributed to the GI's on the line in Vietnam, regularly jammed and was so complicated that it was difficult to clean. The weapon does present "a maintenance problem," the Army's Chief of Research conceded to a Congressional committee.

In a generation the U.S. military has done a complete about-face. On the eve of Hitler's blitzkrieg, U.S. Army generals could not bring themselves to give up the cavalry. An innovator like Billy Mitchell, who could see glorious new missions for his service through aviation technology, was rewarded with a court-martial. Now generals and admirals from the three services scramble over one another to lay claim to the latest, fanciest, and most expensive technology. A technological improvement, even the mere possibility of one, is reason enough to spend millions. Keeping up with "the state of the art" is a sacred obligation. Chairman L. Mendel Rivers of the House Armed Services Committee raged at the memory of McNamara at a Committee hearing for "having the effrontery to try to hold back the state of the art." The former Secretary of Defense didn't understand that he was up against a mystical power for which human reason was a poor match. Quoting the patron saint of the Air Force, General H. H. "Hap" Arnold, Rivers pronounced the first great commandment of unlimited military spending:

> Nobody, but nobody, but nobody can take credit for or stop the state of the art. It is something that will move, and it just pours out of the head of people and it is unconscious and it will go ahead and those who get on the saddle can, and those who can't leave.

A gun may shoot farther or a plane fly faster or a missile land more accurately without any gain in security. The new "capability" may be a source of pride to the service that acquires it, but not a source of real strength. The reason is that in the nuclear age small improvements in performance don't make much difference. It is highly unlikely that the U.S. will ever fight a prolonged conventional war with the Soviet Union in which technological superiority is decisive. The losing side will resort to nuclear weapons long before it concedes victory to the other in a crucial artillery duel or naval battle. The effort to improve the accuracy of the missiles actually decreases security, for it is a signal to the Soviets that the U.S. is developing a "first-strike" force designed to knock out missiles in the Soviet Union. Their reaction, as we have noted, is to aim more missiles at the American people. Matching the Soviets in each weapons system, regardless of whether the system is likely to be used or would have much effect if it were, is a bankrupt and bankrupting policy. Even if the Soviets were putting much effort into germ warfare —and the evidence is that they are not—this would not necessarily mean that we should. Our germs cannot protect us from their germs. Nor is germ warfare any added deterrent. If the Soviets launched a bacteriological attack on the U.S., the President would respond with the sure destruction of nuclear weapons.

Research and development as "insurance" against technological breakthrough is also largely myth. Less than $500 million of the $8 billion research budget goes for research. The rest goes for development—i.e., implementation. Only the research expenditure holds any promise of uncovering "breakthroughs" if by the term we mean fundamental new concepts that might actually change the military balance. A reliable device for locating Polaris submarines and rendering them vulnerable might be an example; fortunately, Admiral Levering

Smith, the developer of the Polaris, states that such a device is technologically unfeasible in the foreseeable future, although one should assume that at some point it may become possible. Fantastic engineering improvements, notably in electronics, have been developed in the postwar period. But except for the brief U.S. nuclear monopoly following World War II, neither side has been in a position to use its laboratory triumphs to blackmail the other.

It has been enough to justify a new weapon by pointing to the fact that the Soviets are building one or *might* build one. The Soviets seem to be much smarter. Because their investment in the technology of death is more modest, they apparently understand that a more "advanced" system does not necessarily enable a military power to do anything more than it could have done with the old system. To the technologists of the Pentagon they appear "conservative." American military planners now need to ask, "What are we going to be able to do tomorrow with the new plane or missile or shell that we can't do today?" The taxpayer is entitled to an answer to the question. Because McNamara introduced the concept of cost-effectiveness into the Pentagon at the technical level, it is now possible to calculate with some accuracy the cheapest way to kill a man under a variety of conditions. But why is there no political definition of "cost" or "effectiveness"? What is the investment of a billion dollars in a new plane really costing us in terms of decaying cities and crushed hopes? Assume that the new weapon will make it possible to kill more people faster, more efficiently, even more comfortably. Is this a desirable goal for America?

II

The Military-Industrial Complex and How It Works

The Economy of Death defies logic. A piece of technology like the ABM is virtually discredited again and again by every former science adviser to the President, a number of Nobel Prize physicists, and several former high officials of the Defense Department itself. Yet the juggernaut moves on. If one rationale for building a new weapons system is exposed as nonsense, others spring up to take its place. The Secretary of State talks about détente and coexistence, and the Secretary of Defense demands the build-up of a first-strike force. The Pentagon demands billions to counter a nonexistent Chinese missile force while ghetto and campus rebellions, police riots, and political assassinations tear away at American society. Why?

The institutions which support the Economy of Death are impervious to ordinary logic or experience because they operate by their own inner logic. Each institutional component of the military-industrial complex has plausible reasons for continuing to exist and expand. Each promotes and protects its own interests and in so doing reinforces the interests of every other. That is what a "complex" is—a set of integrated institutions that act to maximize their collective power. In this chapter we shall look at the various structures of the military-industrial complex to try to understand how and why the decisions are made to allocate our national resources to the Economy of Death.

The defenders of the military establishment like to characterize the growing attacks on the military-industrial complex as conspiracy-mongering. Senator Henry Jackson, one of the staunchest defenders of big military budgets, calls the recent inquiries into military mismanagement and waste the "largest version of the devil theory of history." The injured brass, accustomed to twenty years of nineteen-gun salutes from the public, now charge their critics with the ancient military crime of showing disrespect to officers. When George Mahon, Chairman of the House Appropriations Committee and long-time advocate of big defense budgets, was boorish enough to suggest that perhaps the Navy had been a bit careless to let a $50 million nuclear submarine sink in thirty-five feet of water, L. Mendel Rivers, the Pentagon's most generous friend in Congress, attacked him on the floor of the House for "playing into the hands of the enemies of the military." The enemies he was talking about were not the Russians, but members of the U.S. Senate.

Nothing suggests the existence of a conspiracy more strongly than concerted efforts like these to protect the military establishment from public inquiry and debate. But conspiracy is not the answer. The sad truth is that it is not even necessary. To understand the hold of the Economy of Death on the country, one needs to look at the behavior of institutions, not individuals. To be sure, there are more than a few cases of profiteering, personal enrichment, conflict of interest, and graft. Eisenhower's first Secretary of the Air Force, Harold Talbott, who continued to receive over $400,000 a year from his former company while in office, wrote letters to defense contractors on Air Force stationery suggesting they might like to throw some business to his old firm. The Senate Permanent Subcommittee on Investigations reported in 1964 that numerous companies "pyramided" profits in connection with the missile procurement program. Western Electric, for example, on a contract for "checking over launcher loaders," earned $955,396 on costs totaling $14,293, a respectable profit of 6600 percent. Kennedy's Deputy Secretary of Defense, Roswell Gilpatric, played a major role in awarding the dubious TFX contract to his old client General Dynamics. The postwar successors to the merchants of death, such as Litton, Itek, Thiokol, and LTV, have earned quick fame and fortune. The present Deputy Secretary of Defense, David Packard, for example, parlayed an electronics shop in a garage into a $300 million personal fortune primarily through defense contracts.

A faint odor of corruption pervades the whole military procurement system. (Some early examples can be found in the House Armed Services. Committee Report *Supplemental Hearing Released from Executive Session Relating to Entertainment Furnished by the Martin Company of Baltimore Md. of U.S. Government Officers, September 10, 1959.*) An officer who deals with a defense plant often has access to a variety of personal rewards, including a future with the company. He is likely to eat well, and he need never sleep alone.

But corruption and personal wrongdoing explain very little. In a sense, the managers of the Economy of Death conspire all the time. Men from the services and the defense contractors are constantly putting their heads together to invent ways of spending money for the military. Indeed, that is their job. As John R. Moore, President of North American Rockwell Aerospace and Systems Group, the nation's ninth-ranking

defense contractor, puts it, "A new system usually
starts with a couple of industry and military people
getting together to discuss common problems." Military
officers and weapons-pushers from corporations are "in-
teracting continuously at the engineering level," according to Moore. A former Assistant Secretary of Defense
who has followed the well-trodden path from the Pentagon to a vice-presidency of one of the nation's top military contractors says military procurement is a "seamless
web": "Pressures to spend more . . . come from the
industry selling new weapons ideas . . . and in part
from the military." The problem, then, is not that those
who make up the military-industrial complex act improperly, but that they do exactly what the system expects of them. Corruption is not nearly so serious a
problem as sincerity. Each part of the complex acts in
accordance with its own goals and in so doing reinforces
all the others. The result is a government whose central
activity is planning and carrying out wars. If we look at
the military-industrial complex and how it operates,
it will become clear why it has such a firm hold on
American life and why it cannot be controlled without
major institutional changes.

The Uniformed Military

Let us start our map of the military-industrial complex with the uniformed military. What are their interests? What do they believe?

The modern specialist in violence does not glorify
war. He believes that diplomacy is a polite façade
behind which nations calculate their killing power, and
that political success ultimately depends upon the effec-

tive use of a war machine. Most military men now believe that total war in the nuclear age is not an effective way to use military power. The health of the modern state is not war but preparation for war. The military ethic firmly rejects the idea that the arms race itself could provoke an enemy into war. "They'll fortify the moon if you let them," Churchill once said of the military during the Second World War. Only a few years later a U.S. Air Force general was counseling a Congressional committee on the need to carry the arms race beyond the moon to Venus.

Generals and admirals invariably believe that what is good for the Air Force or the Navy is good for America. A few days after becoming Secretary of the Navy, Paul Nitze discovered a "power vacuum" in the Indian Ocean and a new "requirement" for the fleet. At a Congressional hearing on the B-36, a proposed new bomber for the Air Force, Admiral Arthur Radford denounced nuclear deterrence as "morally reprehensible." It was not until the Navy invented the Polaris submarine-launched nuclear missile that the Admiral decided that the peace of the world depended upon the hydrogen bomb. Each service embellishes "the threat" to serve its bureaucratic interests. The Office of Naval Intelligence is especially good at finding extra Soviet ships which Air Force intelligence always manages to miss. There are tens of thousands of mysterious objects in the Soviet Union which the Army is convinced are tanks but which any Air Force intelligence officer knows are really airplanes.

Each military service has also worked out a view of the world that justifies its own self-proclaimed mission. For the Army, the job is to preserve a "balance of power" and to keep order around the world through counter-insurgency campaigns and limited wars. It should be no surprise that the Air Force view of the world is much more alarmist. "The Soviet leadership is irrevocably committed to the achievement of the ultimate Communist objective, which is annihilation of the capitalist system and establishment of Communist dictatorship over all nations of the world," wrote former SAC Commander General Thomas Powers. According to General Nathan Twining, former Chief of Staff of the Air Force, "the leaders of an organized conspiracy have sworn to destroy America." It is essential to have an enemy worthy of your own weapons and your own war plans. A strategy based on the nuclear annihilation of the Soviet Union is far easier to accept if that country is the embodiment of evil. To rationalize a nuclear arsenal of 11,000 megaton bombs, it is vital to assume that the leaders in the Kremlin are too depraved to be deterred by less. The anti-Communist reflex is the Air Force's biggest political asset.

The preventive-war enthusiasts, vocal as they are in retirement, do not represent the dominant view of the military. True, there was the Air Force general who once exclaimed, "Let's start killing people. People need to respect the United States, and when we start killing people, then there will be more respect for the United States." But most professional military managers prefer to keep preparing for the big war that never comes. There is strong military support in the twentieth century for the view that fighting spoils armies. The Air Force used to define "victory" in nuclear war as having more missiles left than the Russians. It was hard to fire enthusiasm, however, for a war in which the weapons survive and the people die. Even to Air Force officers a full-scale "nuclear exchange" looks like the ultimate disaster in career-planning as well as national policy. This could change, however, if the U.S. continues to build up a "first-strike force" structure. As the Vietnam war illustrates, the military are susceptible to the illusion of victory. The availability of power or even the *appearance* of power tempts men to use it. The following exchange between the Chairman of the Joint Chiefs of Staff and the counsel of a Congressional committee shows how strong are the military man's psychological defenses to the implications of what he proposes:

MR. KENDALL: Suppose the numbers of casualties . . . were doubled [to 160 million for the U.S., 200 million for the Soviets]. . . . Obviously, you would have no country left, neither of us.

MR. WHEELER: Mr. Kendall, I reject the "better Red than dead" theory—lock, stock, and barrel.

But limited war has seemed much more promising. As former Marine Commandant General David Shoup puts it, "War justifies the existence of the establishment, provides experience for the military novice and challenges for the senior officer. Wars and emergencies put the military and their leaders on the front pages and give status and prestige to the professionals." During the early days of the Vietnam war, officers often greeted each other in Pentagon corridors with a standard quip, "It's not much of a war, but it's the only one we've got." In the pre-World War II period the only justification for mobilizing a huge army was to fight a moral crusade against the Kaiser or Hitler. But the permanent mobilization of the postwar period can be rationalized only by a "world responsibility" to be the "Guardian at the Gates."

Military bureaucrats have led exciting but not inordinately dangerous lives acting out the policeman's role. During the biggest pre-Vietnam military operation, the Korean war, the mortality rate of officers holding insurance policies with a leading company was below the average for industry as a whole. The incidence of death in battle fell from 104.4 per 1000 soldiers in the Civil War to 5.5 in the Korean conflict. Despite the heavy casualties among enlisted men and junior officers in Vietnam, the nation has never fought a war in greater comfort or relative safety. The higher officers need not normally confront their own death or anyone else's as part of their professional duties. In the antiseptic world of the Pentagon, captivated by the electronic magic of briefing charts that light up in six colors, instant global communications networks, and the power of Hell at the touch of a button, the top managers of the new constabulary can feel important, safe, and dedicated to a higher cause. "The naval profession is much like the ministry," a Naval captain wrote his son fifty years ago. "You dedicate your life to a purpose . . . You renounce your pursuit of wealth . . . In the final analysis your aims and objects are quite as moral as any minister's because you are not seeking your own good but the ultimate good of your country."

It is not altogether surprising that the military put such a high value on "security" for the nation through the "insurance" of new weapons systems. Military officers have exerted considerable efforts to achieve these goals in their personal lives and careers. The American military establishment is a well-protected, heavily subsidized enclave in American society which offers its constituents the best of the welfare state, including pensions unmatched in industry, complete medical care, travel, consumer bargains, and a steady supply of underlings who must show proper respect under penalty of law. Adventure with low personal risk is an ideal career for a man or a nation.

A very important element of personal security is the belief that your life and your work matter. The profoundly pessimistic view of a world full of enemies which military men invariably share is fundamentally self-serving. Man is evil. Nations are predatory. Therefore the nation must be armed to the teeth. This is not to say that the military man merely pretends to believe his own gloomy vision. He is usually quite serious. What is more, he can find plenty of evidence to support it. Men in charge of nations do use violence to solve political problems and have done so since the beginning of history. That is enough to justify the soldier's role to himself and to society. But the military man never examines his most fundamental assumptions. Challenging an enemy to an arms race is supposed to make him less dangerous and more accommodating, but the evidence is overwhelmingly discouraging, especially in an era when you cannot defeat him without destroying yourself. The effect of the military ideology on foreign policy is disastrous because it excludes all hope of moving international politics out of the Stone Age. In a world where there is no alternative to peace, as General Eisenhower put it, the standard military ideology denies the possibility of peace. "I would hate to see us enter any agreement with anybody," Major General William M. Creasy, the Army's man in charge of stockpiling and thinking up uses for nerve gases and bacterial agents, told a House Committee.

"Do you favor disarmament?" the General was asked.

"Yes, sir. I would like Utopia, too. I just don't believe it is practicable."

Although the armed services of the United States have developed increasingly sophisticated rationalizations for ever more military force, they have not improved on the succinct statement of the German High Command in 1938:

> Despite all attempts to outlaw it, war is still a law of nature which may be challenged but not eliminated. It serves the survival of the race and state or the assurance of its historical future. This high moral purpose gives war its total character and its ethical justification.

The Rise of the Military Establishment

How did the military establishment come to acquire such power in a nation which had a political tradition of condemning large standing armies and in 1938 ranked eighteenth among the nations of the world in land forces? What General Shoup calls the "new militarism" is an outgrowth of the Second World War. The federal government came to play a major managerial role in the economy and to help create and to dispose of a significant share of the national wealth. Within the federal bureaucracy the balance of power shifted decisively to those agencies which handled military power. In 1939 the federal government had about 800,000 civilian employees, about 10 percent of whom worked for national-security agencies. At the end of the war the figure approached 4 million, of whom more than 75 percent were in military-related activities.

Not only did the war radically shift the balance of

power in the federal bureaucracy, catapulting the military establishment from a marginal institution without a constituency to a position of command over the resources of a whole society; it also redefined the traditional tasks of the military. The traditional semantic barriers between "political" and "military" functions were eroded; in the development and execution of strategy, the military were deep in politics. The major decisions of the war, those with the greatest obvious political impact, were made by the President, the Joint Chiefs of Staff, and Harry Hopkins. The Joint Chiefs prepared for diplomatic conferences, negotiated with the Allies. In the war theaters the military commanders, Eisenhower and MacArthur, were supreme. Each obtained the power to pass on all civilians sent to his theater and to censor their dispatches. "Through these controls of overseas communications," the military commentator Walter Millis observed, "JCS was in a position to be informed, forewarned, and therefore forearmed, to a degree no civilian agency could match."

At a time when Stalingrad was still under siege and it would have taken a lively imagination to conjure up a Soviet threat of world domination, United States military planners had already begun planning a huge postwar military machine. As the war ended, the Army demanded a ground force capable of expanding to 4.5 million men within a year. The Navy thought it wanted to keep 600,000 men, 371 major combat ships, 5000 auxiliaries, and a "little air force" of 8000 planes. The Air Force also had specific plans. It wanted to be a separate service and to have a seventy-group force with 400,000 men. With these plans the top military officers made it clear that they were through being fire-fighters called in when the diplomats had failed.

Under the pressure of war, new military instruments for manipulating the politics of other countries had been developed. Those who had put them together argued that the United States would need them in the postwar world, whatever the political environment. Thus the Joint Chiefs of Staff argued successfully for retaining most of the network of bases acquired in the war. The thinking of General William Donovan, the creator of OSS, America's first spy agency, shows the indestructibility of bureaucracies. His assistant Robert H. Alcorn has described his views:

> With the vision that had characterized his development of OSS, General Donovan had, before leaving the organization, made provision for the future of espionage in our country's way of life. Through both government and private means he had indicated the need for a long-range, built-in espionage network. He saw the postwar years as periods of confusion and readjustment affording the perfect opportunity to establish such networks. We were everywhere already, he argued, and it was only wisdom and good policy to dig in, quietly and efficiently, for the long pull. Overseas branches of large corporations, the expanding business picture, the rebuilding of war areas, Government programs for economic, social and health aid to foreign lands, all these were made to order for the infiltration of espionage agents.

A nation that for almost four years had performed stunning managerial feats in moving armies across seas, in producing clouds of airplanes, in training destructive power on an enemy with marvelous efficiency, and, finally, in extracting the abject surrender of two of the leading industrial nations of the world without having enemy soldiers or bombs on its soil or its wealth impaired, was ready to put its confidence in force as the primary instrument of politics. Americans, who had often felt swindled in the dreary game of diplomacy, looked in awe at the immense changes they had wrought in the world with their military power.

To maintain and extend their power in the postwar period, the military have been able to draw on a varied and effective arsenal. The most important weapon has been organization. As we have seen, the military bureaucracies came out of the war with their structures intact. Despite the rapid demobilization of millions of men and the sharp reduction of the defense budget from more than $80 billion to $11.7 billion in the first three postwar years, the institutional relationships of the Economy of Death created in the war were preserved and expanded. In this process the military establishment exploited two other weapons to the fullest: secrecy and fear.

In a bureaucracy knowledge is power. The military establishment has made particularly effective use of its jealously guarded monopoly of information on national-security matters. It has defined the threats, chosen the means to counteract them, and evaluated its own performance. Critics have been disarmed by the classification system and the standard official defense of policy, "If you only knew what I know." Academic consultants who have made their living advising the Department of Defense and writing about national-security affairs have protected their security clearances by discreetly accepting the Pentagon's assumptions. Except for a handful of Quakers, radicals, independent scientists, and incorrigible skeptics, no one during the 1940's and 1950's challenged the growing power of the military. The fact that the Pentagon was assuming a central position in American life was obvious. But the justification seemed equally obvious, and the danger was ignored.

Because of its exclusive hold on top-secret truth, the Pentagon was in a position to scare the public into supporting whatever programs the Administration put forward. The Department of Defense became a Ministry of Fear issuing regular warnings about a highly exaggerated threat of a Soviet attack in Europe and a nuclear strike against the United States long before the Soviets had the means to carry it out. Joseph R. McCarthy was a helpful ally in creating a climate of fear until he turned against the Army in a last suicidal gesture. But McCarthyism preceded McCarthy. Alger Hiss, the old China hands, the Poland losers, the Czechoslavakia losers, and other "vendors of compromise" in the State Department, as Senator John F. Kennedy would later call them, became tabloid celebrities long before Senator McCarthy advertised his "list" of 205 known Communists in the State Department. In this atmosphere anyone who dared to suggest that the country was spending too much money on defense was obviously either a traitor with a plan to leave the country "naked to attack" or a coward who preferred to be red rather than dead.

Military officers constantly held up to the public the specters of Hitler and Pearl Harbor. The only security in a dangerous and irrational world was to run it. On his office wall the first Secretary of Defense, James Forrestal, hung a framed card on which was printed the official lesson of World War II: "We will never have peace until the strongest army and the strongest navy

are in the hands of the world's most powerful nation."
It was now America's turn to be Number One in the
world and to play out the historical role of earlier
empires. It did not matter that the age of empire was
over or that the age of nuclear weapons had come. In
military bureaucracies it is standard procedure to pur-
sue the most promising strategy for preventing the last
war.

The Pentagon and the Public

The Armed Services have developed an elaborate
packaging system to market their principal products to
the public. Fear, of course, has been the biggest seller.
It is packaged in a variety of ways. The National Security Seminar of the Industrial College of the Armed
Forces, for example, has held 293 seminars in 161
cities, attended by more than 175,000 reserve officers
and civilians. America "is faced by the greatest danger
the world has ever known—the cancer of communism,"
Captain William A. Twitchell of the U.S. Navy told an
understandably nervous audience at a typical National
Security Seminar in Sioux Falls, South Dakota, in 1963.
Colonel Charles Caple of the Air Force assured the
people of Sioux Falls that any Soviet leader would have
the same aim: "to destroy our way of life. We cannot
coexist with these people." Representatives of industry,
labor, business, education, religion, the professions,
government, and civic and women's organizations are
encouraged to attend as a patriotic duty.

There are 6140 public-relations men on the Pentagon payroll. The information branch of the Public Affairs office of the Department of Defense alone, with a

budget of $1.6 million, employs more than 200 officers and civilians located in the Pentagon and in key cities around the country. The Office of Information for the Armed Forces has a $5.3 million budget which is used for a global radio network that reaches vast civilian as well as military audiences. The Armed Services Radio and Television Service operates 350 stations in 29 countries and 9 U.S. territories, spends $10 million a year and has 1700 employees. It is the largest broadcasting network in the world. Another agency for merchandising the Pentagon is the Armed Forces Motion Pictures, Publications and Press Service, which not only prints 8.5 million copies a year of some 70 military publications but also produces film clips and tapes for commercial TV and radio stations.

The three services have even larger public-relations activity. In 1946 the services were the nation's third-largest advertiser. The Army supports public-relations officers at each military post, an Army Exhibit Center that travels around the country, and an Army Hometown News Center, operating out of Kansas City, which makes sure that local papers across the country continue to tell the Army story. "The Army Hour" on radio and "The Big Picture," a TV propaganda offering, are carried by several hundred stations each week.

During the Truman, Eisenhower, and early Kennedy years, military officers were encouraged to make speaking tours, alerting the country to the coming Holy War against communism, a war which one day could end in the justifiable homicide of enough communists to kill the idea itself. Lieutenant General Arthur Trudeau, the chief of Army Research and Development, used to like to give speeches about "the horrible maelstrom of World Communism that is sucking our nation into the vortex of death and destruction." He assured his enthusiastic audiences that the Army was continually thinking up new weapons to counter "the toxic darkness of World Communism." Admiral Radford liked to call for "total victory." Air Force generals warned of the Soviet bid for world domination which could be thwarted only by U.S. air power. In part because of the activity of Captain Kenneth J. Sanger, who arranged for fellow officers to give more than 400 blood-and-guts speeches in Seattle in a single year, Senator J. William Fulbright and others were able to stop some of the military war talk. The Department of Defense must now review the speeches of military officers to make sure that the foreign policy of the Air Force does not diverge so much from that of the State Department as to confuse the enemy.

The Pentagon's major campaign to sell the Sentinel ABM came to light when someone leaked the Starbird Memorandum. This document, written by Lieutenant General A. D. Starbird, manager of the Sentinel Program, is a step-by-step outline of a concerted public pitch to convince the taxpayer of the necessity of ABM. It calls for "journalistic tours," briefings for Congressmen, the preparation of magazine articles, and the coordinated use of the considerable communications network already described. An important part of the campaign is covert. "We will be in contact shortly with scientists who are familiar with the Sentinel program and who may see fit to write articles for publication supporting the technical feasibility and operational effectiveness of the Sentinel system," Secretary of the Army Stanley Resor wrote to the Secretary of Defense. Those scientists willing to cooperate in the ABM brainwash will receive "all possible assistance" from the

Army. In addition to using the Pentagon's own resources, General Starbird advises his public-relations
staff to "cooperate and coordinate with industry on
public relations efforts by industries involved in the
Sentinel program." It is rare that individual bureaucracies of the Pentagon are indiscreet enough to reveal so
graphically the ways in which they manipulate public
opinion to justify their expansion. Fortunately, Congress long ago took the first step in dealing with the
problem of bureaucratic brainwashing. The Arms Control and Disarmament Agency is prohibited by law
from spending any funds to propagandize the American people!

Next to fear, pride is the Pentagon's most potent
weapon for conducting psychological warfare in American society. The Community Relations Program of the
Department of Defense specializes in packaging pride.
According to Department of Defense Directive Number
5410.18, issued on February 9, 1968, the following are
proper community-relations activities:

> liaison and cooperation with industrial, technical,
> and trade associations, with labor, and with other
> organizations and their local affiliates at all levels;
> Armed Forces participation in international, na-
> tional, regional, State and local public events; in-
> stallation open houses and tours, embarkations in
> naval vessels, orientation tours for distinguished
> civilians; people-to-people and humanitarian acts;
> cooperation with government officials and commu-
> nity leaders; and encouragement of Armed Forces
> personnel to participate in all appropriate aspects
> of local community life.

Tours of installations, demonstrations of aircraft and
new weapons, and participation in community events

by air rescue teams are designed to let the citizen bask a little in the wonders of military technology. They are a form of psychological rebate for the taxpayer.

As John Swomley has shown in *The Military Establishment,* "orientation conferences" offering ego-boosting holidays for prominent citizens have been an especially effective way for the military to win friends in strategic places. Bennett Cerf, publisher and TV personality, considered it "one of the biggest honors and luckiest breaks of my career" to have been invited by the Secretary of Defense to Fort Benning, Georgia, in 1950 along with other "leading citizens" to witness a display of "remarkable new recoilless weapons (and other arms still considered secret)." Cerf was quite aware that the purpose of all this exciting attention was "to spread the good word." On his return, he was only too happy to comply in the pages of the *Saturday Review of Literature:*

> I came home revitalized and simply bursting to shout from the housetops this deep-felt conviction that when and if a war comes with Russia or anybody else this country is blessed with the basic equipment and leadership to knock the hell out of them. We need more fighter planes and more carriers. We need more men in the Armed Forces. Our intelligence and propaganda departments need bolstering most of all. The money already allotted to defense has been on the whole wisely spent. In the light of day to day new developments, increased appropriations are not only a wise investment but an absolute must.

The Pentagon has another, somewhat more subtle technique for cultivating public pride in the services. From time to time it cooperates in making a Hollywood movie. The film companies get literally millions of dollars of free or almost free sets in the form of loaned aircraft carriers and military installations. The services get millions of dollars' worth of free publicity. Recently Twentieth Century-Fox, for a film called *Tora! Tora! Tora!,* hired the aircraft carrier *Yorktown* for a few days, dressed it up as a Japanese aircraft carrier, and used it to re-enact the bombing of Pearl Harbor. Ten U.S. ships on active duty were employed for the extravaganza. Two civilian pilots died during the filming, and a number of off-duty servicemen were burned. The Navy public-relations office which handled the affair took pains to point out that the men dressed up in Japanese uniforms on the *Yorktown*'s deck were not sailors but marines. Navy officials admitted to reporters that they had been aware of certain risks connected with the enterprise, but concluded that the movie's message justified the undertaking. The history of the great days of the aircraft carrier was bound to be particularly inspiring to the Navy in a year in which it was seeking funds to build fifteen more of them.

The Militarized Civilians

"The country is looking for a scapegoat. First it was the draft, then recruiters, then Dow Chemical, and now it's the bloody generals," Major General Melvin Zais, Commander of the 101st Airborne Division in Vietnam, complained to an interviewer from *Time*. Many military officers view the belated but growing concern in Congress over uncontrolled military expenditure as an attack on the uniform. In a sense, it is. When General William Westmoreland appeared in full regalia before Congress to make claims about the Vietnam war which no one with a working television set could believe, the credibility gap assumed cavernous proportions. For the first time in a generation, the leaders of the military establishment have been challenged to produce facts and rational arguments to justify their claim to the biggest bite of the tax dollar. Credentials alone are no longer enough. Neither, one hopes, are the traditional national-security slogans about the Soviet Threat or the Chinese Threat, no matter how blood-curdling the rhetoric. Foolishness and waste in the Pentagon, the inevitable by-products of any institution with too much money to spend, are finally under attack. A patriotic American can only hope the attack will grow.

Nevertheless, the uniformed military are not the primary target of a serious political effort to shift from the Economy of Death to the Economy of Life. The principal militarists in America wear three-button suits. They are civilians in everything but outlook. Not the generals but the National Security Managers—the politicians, businessmen, and civil servants who rotate through the paneled offices of the Pentagon, the State Department, the Central Intelligence Agency, the Atomic Energy Commission, and the White House—have been in charge of national-security policy.

In the postwar years there have been, to be sure, some challenges to the traditional American principle of civilian control of the military. General MacArthur denounced as a "dangerous concept" the idea that members of the armed forces owe primary allegiance and loyalty "to those who temporarily exercise the authority of the executive branch of government." The President may be the Commander-in-Chief under the Constitution, but it is "the country," the General asserted, which the soldier must swear to serve. Yet President Truman was able to secure the unanimous assistance of the Joint Chiefs of Staff in removing Mac-Arthur for insubordination in the conduct of the Korean war, and as the old soldier faded away, he became an advocate of nuclear disarmament. In the late Eisenhower years, according to Professor Samuel Huntington's studies, the Secretary of Defense rejected only four out of 2954 unanimous recommendations of the Joint Chiefs, yet many prominent generals resigned publicly in protest against Ike's budget-cutting. The top brass reluctantly went along with Robert McNamara's decisions to cut back on certain weapons systems favored by the Air Force, but they sniped at him in the press and, particularly in his final year at the Pentagon, grew bold in public criticism of the Secretary before the Armed Services Committees. Generals on active duty supplied information to right-wing groups like the American Security Council to assist them in their concerted campaign to discredit the Secretary. As Richard Goodwin, Special Assistant to Presidents Kennedy and Johnson, has pointed out, McNamara took office "with the avowed aim of establishing greater civilian control" over the military:

> Yet, the harsh fact of the matter is that when he left, the military had greater influence over American policy than at any time in our peacetime history. In the name of efficiency we unified many of the operations of the armed services, encouraged greater intimacy between the military and industry, and instituted the deceptive techniques of modern computer management, realistic or hardheaded, to solve problems and invest money and

> use power unguided by ultimate aims and values.
> . . . You can ask a computer whether you have
> the military capacity to accomplish an objective. It
> will answer either "yes" or "no." It will never say,
> "Yes, but it is not a good idea."

By the end of the Johnson Administration, the uniformed military had acquired considerable independent
political power. They had powerful friends in Congress.
Reserve commissions were held by 139 members, and
several of these were generals. The Joint Chiefs had
their terms of office extended from two to four years,
which meant that a new President had to fire the nation's top military if he wanted to appoint his own men.
A retired general ran for the Senate in New Hampshire,
and General Curtis LeMay campaigned as George
Wallace's running mate in the Presidential race of
1968. Both argued that the civilian leaders in the Pentagon were selling out the country. These were straws in
the wind signifying the increasing frustration and anger
of the military against civilian authorities who ordered
them into war but would not let them win it. But the
public was unimpressed. No man on horseback has yet
emerged who appears able to challenge the civilian
leadership in an election, and the supply of marketable
war heroes is dwindling.

President Kennedy once confided to a close friend
whom he had appointed Under Secretary of the Navy
that a military coup along the lines of the one described
in the popular novel *Seven Days in May* could happen
in his Administration if, for example, the Bay of Pigs
fiasco were ever repeated. Despite the frequency of
military coups in the modern world, a seizure of government by the military in the United States still seems

remote. But surrounding the White House with tanks is only one way to militarize the country. The Joint Chiefs of Staff possess sufficient power today so that the President of the United States cannot simply order them. He must negotiate with them. For example, as Elizabeth Drew of the *Atlantic* has revealed, the Joint Chiefs of Staff exacted a price from President Johnson for their agreement to support the cessation of bombing over North Vietnam. They insisted upon taking the bombs they had counted on dropping on the North and dumping them on the South and Laos. There is a deep fear pervading the civilian leadership that the military, if sufficiently provoked, might pit their professional credentials against the "politicians" in a public confrontation.

Thus civilian control of the military has been maintained throughout the long years of the Cold War, but the price has been the militarization of the civilian leadership. Generals and admirals continue to take orders from the President as Commander-in-Chief, but the President spends about 90 percent of his time building and, to use the State Department term, "projecting" America's military power. Increasingly, the civilian managers have come to see the world through military eyes. They plan for "greater than expected threats." They avoid arms-control agreements that could "degrade" our military forces. They think a new "option" and a new airplane are the same thing. They seek "prestige" through intimidation of the weak.

It was a preacher's son from Wall Street, John Foster Dulles, who as Secretary of State spent most of his time building military alliances with more than forty nations. It was an economics professor from M.I.T., Walt Rostow, who made the earliest and most vigorous case for solving the problem of South Vietnam by bombing North Vietnam. It was an investment banker, James Forrestal, who designed the National Security Council without machinery to balance military requirements and domestic needs. It was a Washington lawyer, Dean Acheson, who thought that negotiating the Berlin crisis of 1961 instead of staging a military confrontation with a high risk of nuclear war was "weakness." It was a Massachusetts politican, John F. Kennedy, who was prepared to risk a minimum of 150 million lives to face down Khrushchev in the Cuban missile crisis of 1962. The instinctive approach to international conflict has been to reach for a gun.

Indeed, there is considerable evidence that the civilian managers, particularly at the beginning of the postwar period, have been far readier than the military to commit American forces to actual combat. Apprenticed to the military in World War II, the top civilian national-security elite absorbed the basic military outlook but not the soldier's professional caution. Perhaps because they lacked combat experience, they underestimated the difficulties and risks in using military power. In 1946 the Joint Chiefs of Staff cooled the State Department's enthusiasm for sending an ultimatum to Yugoslavia for shooting down an American plane. In the earliest days of the Cold War it was the State Department that kept urging a big military build-up to furnish "support for our political position," while the Defense Department set more modest goals for itself. Secretary of State Dean Acheson, not the Joint Chiefs of Staff, made the first recommendations to commit U.S. military forces to repel the Korean invasion. General Matthew Ridgway, Chief of Staff of the Army, opposed John Foster Dulles' proposal to intervene militarily in Indo-China in 1954.

The Joint Chiefs opposed Walt Rostow's plan to invade Laos in 1961. The military did not recommend commitment of forces either to aid the Hungarian revolution or to tear down the Berlin Wall. When David Lilienthal, the civilian chairman of the Atomic Energy Commission, asked the Joint Chiefs of Staff in 1949 what uses they would have for a hydrogen bomb, they couldn't think of one. The civilians directed the military to think harder.

The militarized civilians have surpassed the military themselves in embracing a "realism" which envisages no alternative to an escalating arms race but annihilation, perpetuates an aimless war by calling it a commitment, and measures the nation's greatness in megatons. Far more responsive to the bureaucratic interests of the services than to the wider political interests of the American people, they believe that America's principal role in the world is to acquire more power. Indeed, they see the accumulation of power as an end in itself. Thus it is they, rather than the uniformed military, who must assume responsibility for the distortion of national priorities. The armed services always want a more perfect fighting machine, but the civilian leadership does not have to give it to them. The military traditionally look upon the population as a mobilization base for the armed forces, forgetting that the army exists to serve the people and not the other way around. But the civilians at the top need not share that view.

Yet this is precisely what has happened. Civilian strategists and diplomats have developed a strategy that has made the American people instruments of political warfare. President Kennedy in 1962 was ready to risk the nuclear destruction of American society rather than negotiate with the Soviets about removing missiles no closer to our country than the U.S. missiles in Turkey were to theirs. To have bargained instead of threatened would have cost little. Indeed the decision to remove the Turkish missiles had been made months before the confrontation. "Your Daddy may have started World War III," Lyndon Johnson told his daughter on the day he accepted the advice of Dean Rusk, McGeorge Bundy, Walt Rostow, and Robert McNamara, and risked war with the Soviet Union and China by bombing their ally, North Vietnam. The side willing to take the greater risks with the lives of its own people wins this global version of the game of "chicken."

The temptation to try to solve political problems through violence is almost irresistible for the political leaders of a great power because the risk of retaliation seems low and the military man's "solutions" offer the illusion of toughness, practicality, and certitude. Factors which can be fed into computers such as "kill ratios" sound more persuasive than political analysis, which is hard to prepare and hard to comprehend. To understand the true interests of the American people in a remote area of the world, a diplomat must not only try to understand who the Vietnamese are and what they want, but must continually try to rediscover America's own interests. It is much easier to avoid both processes by treating the outside world as a collection of statistics, nuisances, and threats, ignoring domestic needs and treating international politics as gambling in a good cause. Arthur Schlesinger, Jr.'s account of the deliberations preceding the Bay of Pigs illustrates the point:

> The advocates of the adventure had a rhetorical advantage. They could strike virile poses and talk of tangible things—fire power, air strikes, landing

craft, and so on. To oppose the plan, one had to invoke intangibles—the moral position of the United States, the reputation of the President, the response of the United Nations, "world public opinion" and other such odious concepts. These matters were as much the institutional concern of the State Department as military hardware was of Defense. . . . I could not help feeling that the desire to prove to the CIA and the Joint Chiefs that they were not soft-headed idealists but were really tough guys too influenced State's representatives at the Cabinet table.

Whatever the human dynamics that may explain it, the consequences of deferring to the military outlook have been disastrous. In recent years, under the tutelage of their civilian superiors, the military have managed to lose most of their earlier reticence. In 1962 the Joint Chiefs of Staff favored immediate bombing and invasion of Cuba. In Vietnam they have authorized the annihilation of population centers and urged the bombing of dikes and flooding as a strategy for winning the war. The abdication by the civilian leadership of their responsibility to find political rather than military solutions to the problems of twentieth-century man has encouraged the men in uniform to play an increasingly important and hawkish role in national policy.

The National Security Managers and the National Interest

Who are the key civilian foreign-policy decision-makers? How does one get to be a National Security Manager? Why do they think as they do? The questions are important, for the interests and beliefs which these men bring to their high office decide for the rest of us what the national interest is. "Foreign policies are not built on abstractions. They are the result of practical conceptions of national interest," Charles Evans Hughes noted when he was Secretary of State. The key word is *practical*. One man's Utopia is another man's Hell. Like the flag, the national interest can mean many different things to different people. In times past, those who managed the affairs of a great nation invariably defined the national interest in terms of power and glory. Whatever accrued to the majesty of the state was in its interest. It did not matter how much the people had to be taxed to pay for it or how many had to die. Anything that authorities decided was necessary to the health of the state was, by definition, in the national interest. President Johnson invoked the national interest to justify sending 600,000 troops to Vietnam and keeping them there despite the disastrous domestic consequences. In postwar America there has been a twenty-five-year consensus on the national interest. There is money for weapons, but not for people. Social decay must be accepted as the price of power. Security is to be achieved by preparing for the worst in meeting foreign threats and assuming that the crisis of our own society will take care of itself.

Who has decided that this is what the national interest is all about?

Since 1940 about 400 individuals have held the top civilian national-security positions. These men have defined the threats for the nation, made the commitments that were supposed to meet these threats, and determined the size of the armed forces. They have been above electoral politics. With few exceptions, the men who have designed the bipartisan foreign policy have never held elective office. Their skills have not been those of the politician, who must at least give the appearance of solving problems or reconciling competing interests if he hopes to get re-elected, but those of the crisis-manager. Dean Rusk characterized his personal goal in office as Secretary of State as handing the Berlin crisis over to his successor in no worse shape than he found it. This is the managerial or "keep the balls bouncing" view of statecraft characteristic of those who count on being somewhere else when the ball drops.

If we take a look at the men who have held the very top positions, the Secretaries and Under Secretaries of State and Defense, the Secretaries of the three services, the Chairman of the Atomic Energy Commission, and the Director of the CIA, we find that out of ninety-one individuals who held these offices during the period 1940–1967, seventy of them were from the ranks of big business or high finance, including eight out of ten Secretaries of Defense, seven out of eight Secretaries of the Air Force, every Secretary of the Navy, eight out of nine Secretaries of the Army, every Deputy Secretary of Defense, three out of five Directors of the CIA, and three out of five Chairmen of the Atomic Energy Commission.

The historian Gabriel Kolko investigated 234 top foreign-policy decision-makers and found that "men who came from big business, investment and law held 59.6 percent of the posts." The Brookings Institution volume *Men Who Govern,* a comprehensive study of the top federal bureaucracy from 1933 to 1965, reveals that before coming to work in the Pentagon, 86 percent of the Secretaries of the Army, Navy, and Air Force were either businessmen or lawyers (usually with a business practice). In the Kennedy Administration 20 percent of all civilian executives in defense-related agencies came from defense contractors. Defining the national interest and protecting national security are the proper province of business. Indeed, as President Coolidge used to say, the business of America is business.

In *Democracy in America,* Alexis de Tocqueville worried that the United States might not be successful in its foreign relations because "foreign politics demand scarcely any of those qualities which a democracy possesses; and they require on the contrary the perfect use of almost all those faculties in which it is deficient." He said that an aristocracy was better for running foreign policy because a government of the few could keep secrets, was invulnerable to the passions of the mob, and knew how to exercise great patience. Tocqueville would be greatly reassured by the way the conduct of foreign policy has evolved in America, for the National Security Managers exercise the power to make life-and-death decisions with very little interference from the rest of us.

They also constitute a social elite. At least a quarter of them are members of Washington's Metropolitan Club, a fair imitation of a London gentlemen's club. William Domhoff has noted that since 1944 five out of the seven Secretaries of State were listed in the *Social Register.* So also were many of the key figures in the War and Defense Departments, including Henry Stimson, Robert Patterson, John McCloy, Robert Lovett, James V. Forrestal, Thomas Gates, and Neil McElroy. The CIA, which is an outgrowth of the wartime OSS, has also put considerable emphasis on "background."

(Unkind critics of OSS used to say the initials stood for "Oh So Social," so pervasive was the old-school-tie character of America's first spy agency.) The OSS roster included the most famous upper-class names in the country, including Paul Mellon, David Bruce, Albert DuPont, Junius S. Morgan, Lester Armour, and Lloyd Cabot Briggs. In more recent years Allen Dulles, William Bundy, Robert Amory, and John McCone have continued the tradition of bringing good breeding to the practice of espionage.

The collection of investment bankers and legal advisers to big business who designed the national-security bureaucracies and helped to run them for a generation came to Washington in 1940. Dr. New Deal was dead, President Roosevelt announced, and Dr. Win-the-War had come to take his place. Two men—Henry L. Stimson, Hoover's Secretary of State and a leading member of the Wall Street bar, and James V. Forrestal, president of Dillon Read Co., one of the biggest investment bankers—were responsible for recruiting many of their old friends and associates to run the war. In the formative postwar years of the Truman Administration, when the essential elements of U.S. foreign and military policy were laid down, these recruits continued to act as the nation's top National Security Managers. Dean Acheson, James V. Forrestal, Robert Lovett, John McCloy, Averell Harriman, all of whom had become acquainted with foreign policy through running a war, played the crucial roles in deciding how to use America's power in peace.

Once again it was quite natural to look to their own associates, each an American success story, to carry on with the management of the nation's military power. Thus, for example, Forrestal's firm, Dillon Read, contributed Paul Nitze, who headed the State Department Policy Planning Staff in the Truman Administration and ran the Defense Department as deputy to Clark Clifford in the closing year of the Johnson Administration. William Draper, an architect of U.S. postwar policy toward Germany and Japan, came from the same firm. In the Truman years twenty-two key posts in the State Department, ten in the Defense Department, and five key national-security positions in other agencies were held by bankers who were either Republicans or without party affiliation. As Professor Samuel Huntington has pointed out in his study *The Soldier and the State,* "they possessed all the inherent and real conservatism of the banking breed." Having built their business careers on their judicious management of risk, they now became expert in the management of crisis. Their interests lay in making the system function smoothly—conserving and expanding America's power. They were neither innovators nor problem-solvers. Convinced from their encounter with Hitler that force is the only thing that pays off in international relations, they all operated on the assumption that the endless stockpiling of weapons was the price of safety.

The Eisenhower Administration tended to recruit its National Security Managers from the top manufacturing corporations rather than from the investment banking houses. To be sure, bankers were not exactly unwelcome in the Eisenhower years. Robert Cutler, twice the President's Special Assistant for National Security Affairs, was chairman of the board of the Old Colony Trust Company in Boston; Joseph Dodge, the influential director of the Bureau of the Budget, was a Detroit banker; Douglas Dillon, of Dillon Read, was Under Secretary of State for Economic Affairs; Thomas Gates,

the last Eisenhower Secretary of Defense, was a Phila-
delphia banker and subsequently head of the Morgan
Guaranty Trust Company.

But most of the principal figures of the era were
associated with the leading industrial corporations, either as chief executives or directors; many of these
corporations ranked among the top 100 defense contractors. Eisenhower's first Secretary of Defense was
Charles Wilson, president of General Motors; his second was Neil McElroy, a public-relations specialist who
became president of Procter and Gamble. One Deputy
Secretary of Defense was Robert B. Anderson, a Texas
oilman. Another was Roger Kyes, another General Motors executive, and a third was Donald Quarles of
Westinghouse.

In the Eisenhower Administration two fundamentally
different views of defense spending and national security clashed. The advocates of each were National Security Managers with business backgrounds. Nelson
Rockefeller, who had invested in such defense companies as Itek, McDonnell Aircraft, and Thiokol, was also
a leading advocate of bigger military budgets. In the
mid-1950's he and his brothers commissioned a report
on *International Security—The Military Aspect,*
drafted by Henry A. Kissinger, at present President Nixon's Special Assistant for National Security Affairs. The
report called for a steady rise of $3 billion a year in the
defense budget, and a follow-up report stated that the
country should be prepared to spend $70 billion a year
by 1967. (These recommendations, though slightly on
the low side, turned out to be remarkably accurate
projections.) At about the same time the Gaither Report, chaired jointly by H. Rowland Gaither, the man
who was instrumental in setting up the Rand Corpora-

tion, the Air Force "think tank," and Robert C. Sprague, head of a Massachusetts military-electronics firm, called for "substantially increased expenditures" for national security including a $22 billion fallout-shelter program. Robert Lovett and John J. McCloy appeared before a meeting of the National Security Council on November 7, 1957, to assure the President that the financial community would support the steep increase in defense costs.

But Secretary of the Treasury George Humphrey, arguing that the Soviet strategy was to "make us spend ourselves into bankruptcy," refused to go along with the Gaither Committee recommendations. "You know how much money you're asking for?" the Secretary asked the staff director of the Gaither Committee. Pointing to the Washington Monument, which was clearly visible from his window, Humphrey estimated that if the amount of money requested were paid in $1000 bills, the pile would reach up to the top of the monument and fifty-six feet beyond. "As long as I'm here, you're not getting one of those bills." President Eisenhower went along with the fiscal conservatives and kept the budget below what Rockefeller and the defense contractors wanted, thus exposing his Administration to the "missile gap" charges of the 1960 campaign. In 1962 the former President publicly declared that "the defense budget should be substantially reduced."

The most influential of the National Security Managers were able to set the tone of national policy for a generation by combining perseverance and longevity. In 1938 Adolph Berle ran Latin American policy for F.D.R. In 1961 he designed the Alliance for Progress for J.F.K. In 1940 Dean Acheson was drafted into service to prepare the legal justification for giving Britain some old destroyers. In 1968 he was still a close (and hawkish) adviser to the White House. In 1941 Robert Lovett was Assistant Secretary of War in charge of the Air Corps. In the Truman Administration he served as Number Two in the State Department and the Defense Department. In 1961 President Kennedy offered him his choice of the three top Cabinet posts. He turned them all down, but left his enduring mark on American history by recommending his old assistant Dean Rusk. Rusk's service in the State Department spanned the Truman, Kennedy, and Johnson years. Paul Nitze, Averell Harriman, and Clark Clifford all occupied influential positions at the beginning of the Truman Administration and at the end of the Johnson Administration twenty-two years later.

When President Kennedy was elected on his campaign promise to "get the country moving again," the first thing he did was to reach back eight years for advice on national security. Many of the men he appointed as top National Security Managers of the New Frontier were the old faces of the Truman Administration. In addition to Dean Rusk, a strong MacArthur supporter in the Korean war who wrote the memorandum urging that U.N. forces cross the 38th Parallel in Korea, Kennedy's State Department appointments also included George McGhee, a successful oil prospector and principal architect of Truman's Middle East policy. Adolph Berle, Averell Harriman, Paul Nitze, John McCone, John McCloy, William C. Foster, and other experienced hands from the national-security world also made it clear that the new Administration would follow familiar patterns.

However, the Kennedy Administration brought in some new faces, too. McGeorge Bundy, the Dean of the

Faculty of Arts and Sciences at Harvard when Kennedy was an overseer, impressed the new President with his crisp, concise, and conventional analysis. The son of Henry Stimson's close wartime assistant, the Boston trust lawyer Harvey H. Bundy, McGeorge was a publicist who had put the Stimson papers in order and published a highly laudatory edition of Acheson's speeches while the Secretary (the father-in-law of Bundy's brother William) was under attack in the Congress. The teacher of a highly popular undergraduate course on power and international relations, Bundy appeared to have all the necessary qualifications to be the Special Assistant for National Security Affairs. Walt Rostow, the energetic M.I.T. professor who, in his own words, had not spent a year outside of the government since 1946, was promoted to a level appropriate to a man who had just published a book totally demolishing Marx. He was made Bundy's assistant.

Into the Defense Department also came a contingent of systems analysts and nuclear strategists from the Rand Corporation. These were not exactly new faces either. During the Eisenhower years they had been in and out of Washington, prodding the military services into more innovative thinking, which usually meant buying new hardware. Now they became the bosses of the men they had been advising. Charles Hitch, the Rand economist who had written a book on cost effectiveness in defense spending, was invited to put his program-packaging notions into effect as comptroller of the Defense Department. With Hitch came Alain Enthoven, Harry Rowan, and other Rand alumni, who became known as the "whiz kids."

However, the principal positions, with the exception of Bundy's, were filled according to the old patterns.

Robert McNamara was president of Ford instead of General Motors; Roswell Gilpatric, appointed Deputy Secretary of Defense, was a partner of a leading Wall Street firm. In career experience he differed from his immediate Republican predecessor, James Douglas, in three principal respects. His law office was in New York rather than Chicago. He had been Under Secretary of the Air Force, while Douglas had been Secretary. He was a director of Eastern Airlines instead of American Airlines.

The Johnson Administration continued most of the Kennedy national-security appointments. Perhaps because of his generally bad relations with the academic community, Johnson appointed more professors to be National Security Managers than any of his predecessors. Lincoln Gordon and Covey Oliver, from the Harvard Business School and the University of Pennsylvania Law School, respectively, were made Assistant Secretaries of State in charge of Latin America. Although neither was a neophyte in government service and Gordon had a long history of business connections in Latin America, their appointments were departures from the Roosevelt and Truman practice of appointing the largest American investors in Latin America, notably Nelson Rockefeller and Spruille Braden, to oversee U.S. policy there.

The Nixon Administration brought in a veteran academic foreign-policy analyst, two business lawyers, one of the leading defenders of the military in the House, and one of California's most successful defense contractors to decide the national interest. Henry Kissinger, the academic, had made his reputation by advocating the judicious use of so-called "tactical" nuclear weapons on the battlefield. He then enhanced it by retracting his inherently mad proposal, thus demonstrating flexibility. The business lawyers, William Rogers and Elliot Richardson, had had no previous national-security or foreign-policy experience, which was no doubt an advantage. (Richardson's immediate prior experience had been in Massachusetts politics.) The Congressman, Melvin Laird, had an uncanny faculty for seeing all sorts of gaps in America's "security posture" and had written a book recommending that the U.S. launch a nuclear strike against the Soviet Union if "the communist empire further moves to threaten the peace."

The National Security Managers, like the uniformed military, have looked at the world through very special lenses, and the result has been a remarkable consensus. The policy toward NATO and German rearmament laid down by Dean Acheson in the late 1940's, the policy toward Vietnam laid down by John Foster Dulles in the early 1950's, and the policy of rearmament and nuclear strategy developed by the Truman Administration continue to this day without having been subjected to serious re-examination in five Presidencies. Most of the men who have set the framework of America's national-security policy, as I found when I studied the background of the top 400 decision-makers, have come from executive suites and law offices within shouting distance of one another in fifteen city blocks in New York, Washington, Detroit, Chicago, and Boston. It is not surprising that they emerge from homogeneous backgrounds and virtually identical careers with a standard way of looking at the world. They may argue with one another about means but not about ends. It has apparently never occurred to one of them to question seriously the basic assumptions of national-security policy.

The years of American supremacy have meant wealth, fame, comfort, excitement, and a sense of accomplishment for those who have operated at the top of the society. It is hardly surprising that those who have prospered equate the national interest and the status quo. When the term is stripped of geopolitical metaphor and ideological gloss, national security means nothing more complicated than making sure that the American Way of Life continues undisturbed by foreign challengers. But the American Way of Life means very different things to different people. To a Mississippi tenant farmer trying to keep out of his white neighbor's way and to eke out a subsistence living, the American Way of Life leaves something to be desired. Scaring the Russians or the Chinese with an extra supply of missiles is not a first-priority concern. By the same token, it is difficult to see how the real security interests of a family battling rats in the ghetto or of a suburban housewife too fearful to venture out on the city streets connect with the measures that the National Security Managers are taking in their behalf.

Talleyrand once said that France had her conquests and Napoleon had his. In America the suggestion that the views of her leaders are colored by personal or class interests is generally dismissed as character-assassination or Marxism or both. Men who run for office seek "power," but the watchwords of the National Security Managers are "service" and "responsibility." It is easy to debunk the rhetoric of the National Security Managers, who have managed to concentrate greater power of life and death in their hands than Genghis Khan and to increase their fortunes at the same time. But to do so is to miss an important point. The patrician warriors are not cynics but complacent idealists who believe that

America can ultimately enrich the world as she enriches them and their friends.

The National Security Manager is likely to be among the most shrewd, most energetic, and often most engaging men in America, but he has little feel for what is happening or should happen in his country. When Charles Wilson, the former president of General Motors who became Eisenhower's Secretary of Defense, blurted out his delightful aphorism, "What is good for the country is good for General Motors and vice-versa," he was merely restating the basic national-security premise. Using national power, including military force if necessary, to create a "good business climate" at home and abroad is more important than a good social climate in lower-class or middle-class America. An excited stock market rather than the depletion rate of the nation's human and natural resources is the index of America's progress. The National Security Manager sees his country from a rather special angle because, despite the thousands of miles he has logged, he is something of an émigré in his own society. The familiar itinerary takes him from one air-conditioned room to another across the continent—his office on Wall Street or LaSalle Street, his temporary office at the Pentagon or in the White House basement, a visit to a London or a Texas client, a dinner at the Council on Foreign Relations, and so forth. The closest he ever comes to seeing a hungry American is when his dinner companion is treated to bad service at one of Manhattan's declining French restaurants.

Under the stimulus of defense spending, the American economy has boomed, but its benefits have not been equitably shared. The disparity between rich and poor in America has widened. The National Security

Managers have not regarded the redistribution of wealth as a priority concern, for they have had neither the experience nor the incentive to understand the problems of the poor. Their professional and personal interests are with the business and commercial interests which they serve and with which they identify. For a National Security Manager recruited from the world of business, there are no other important constituencies to which he feels a need to respond.

When planning a decision on defense policy, he does not solicit the views of civil-rights leaders, farmers, laborers, mayors, artists, or small businessmen. Nor do people from these areas of national life become National Security Managers. Indeed, when Martin Luther King expressed opposition to the Vietnam war, he was told that it was "inappropriate" for someone in the civil-rights movement to voice his views on foreign policy. The opinions which the National Security Manager values are those of his friends and colleagues. They have power, which is often an acceptable substitute for judgment, and since they view the world much as he does, they must be right. They are also the men with whom he will most likely have to deal when he lays down the burdens of office. "What will my friends on Wall Street say?" the Director of the Arms Control and Disarmament Agency once exclaimed when asked to endorse a disarmament proposal that would limit the future production of missiles.

The idea that spending one's life in the securities market is an essential qualification for dealing in national security is a confusion. There is no reason why the National Security Managers should not represent diverse interests, backgrounds, and ways of looking at the national interest. It is almost unbelievable that of the 400 top decision-makers who have assumed the responsibility for the survival of the species, only one has been a woman.

Conflict of Interest and the National Interest

"One must believe in the long-term threat," James J. Ling, president of Ling-Temco-Vought, recently insisted in an interview with the Washington *Post*. For Ling, who built a $3000 investment in an electrical shop in Dallas into a $3.2 billion conglomerate and the eighth-largest defense contractor, the threat is the equivalent of cash. LTV Aerospace is planning on more than doubling its military sales by 1973 and expects to continue to do about 80 percent of its business with the government in military and space contracts. Like Ling, John R. Moore of North American Rockwell, the ninth-ranking military contractor, believes that "defense spending has to increase in our area." The manager of the Fort Worth Division of General Dynamics, Richard E. Adams, which produced the TFX, is confident that his company will be "little impacted by the cessation of hostilities," which is another way of saying he doesn't need a war to sell hardware. Another aerospace-company official notes that "people are pressing for new programs more intensively than ever."

Much of the pressure for new weapons systems comes from the 100 largest defense contractors, which in 1968 had 67.4 percent of all the business. As J. K. Galbraith has pointed out, a dozen firms that are almost wholly dependent upon military contracts, such as General Dynamics, McDonnell Douglas, Lockheed, and United Aircraft, together with two giants, AT&T and

General Electric, had a third of all Department of Defense contracts in 1968. The weapons-producers are in the business of selling threat-removers, and this requires selling threats as well. Weapons-pushing is an essential activity. In a period of a little over a year (1958–59) the Navy alone received 486 unsolicited bids for new hardware. "We try to foresee the requirements the military is going to have three years off," John W. Bessire, a General Dynamics official, explains. "We work with their requirements people and therefore get new business." Many defense contractors operate "think tanks" that perform contract studies on the "threat environment" for the services to which they also sell hardware. McDonnell Douglas produced one such effort with a less opaque title than most. It was called "Pax Americana."

In the early days of the Cold War, corporations were particularly zealous in public-service advertising, education, agitation, and other tax-deductible activities. The President's Air Policy Commission Report, known as the Finletter Report, based on the testimony of military officers and aircraft and airline corporation executives, presented a terrifying picture of a Russian menace which could be countered only by a seventy-group Air Force capable of visiting "utmost violence" on the Soviet Union. "Whether we like it or not," the Report added, "the health of the aircraft industry, over the next few years at least, is dependent largely upon financial support from government in the form of orders for military aircraft." On March 16, 1948, the United Aircraft Corporation took a full-page advertisement in *The New York Times* to announce the results of the Finletter Report and the Congressional Aviation Policy Board that followed it:

AIR POWER
—to discourage aggression
—to preserve national security
—to promote the total economic and social welfare

A number of other defense companies, according to the studies of Professor Alan Westin, have spent money to promote policies that depend on the ever greater use of force. The American Security Council, according to its brochure, is "an organization through which the private sector of society might utilize its talents and resources in helping meet the Communist challenge to peace and freedom" and to develop "new and original approaches in countering Communism's war to conquer the world." It operates on an annual budget of about $2.3 million, distributes a comic-book condensation of General Thomas Power's *Design for Survival* ("The Communist threat represents . . . the most insidious and gigantic plot in history") and a paperback book entitled *Peace and Freedom Through Cold War Victory,* and conducts a five-day-per-week radio program carried over 1000 stations moderated by a former right-wing Congressman, Walter Judd. About 20 percent of the contributions to the Council, according to John Lewis, chief of the Council's Washington bureau, comes from defense corporations.

Men from defense industry also act as consultants to the Department of Defense, lending their judgment on a variety of politico-military questions. Thus, for example, Richard Montgomery of Boeing, the sixth-largest contractor in 1967, chaired a government panel on the Chinese Threat. Defense contractors play an even more significant role in contributing members to various scientific and advisory boards of the Department of Defense dealing with weapons development and procurement matters. The Air Force Science Advisory Board includes J. Russel Clark, vice-president of Ling-Temco-Vought; William L. Gray, a top engineer from Boeing; Gilbert W. King, vice-president of Itek; and several representatives from other top defense firms. The rest are recruited from the weapons-research laboratories of America's major institutions of learning, including Princeton, Columbia, Johns Hopkins, M.I.T., and Stanford. On the Army Scientific Advisory Committee are men from General Electric, Boeing, and Litton, while the Navy has the Chief Scientist of Ryan Aeronautical Company and the Vice President for Engineering of the General Dynamics Corporation on its Research Advisory Committee. The purpose of these committees is to alert the Pentagon to new scientific discoveries that could be turned into weapons and to suggest individuals and firms to work on their research and development.

The National Security Industrial Association, founded by James V. Forrestal in 1944 to make sure that "American business will stay close to the services," is, according to a Forrestal quote on the masthead of its monthly newsletter, an organization "of plain American citizens who are interested in the security of the United States." To qualify for membership, however, a plain American must also be an executive of a defense contractor. The purpose of the organization is to lobby the Department of Defense to adopt or to continue practices that benefit the member corporations. Thus, for example, in the annual report for 1964, the President, R. E. Beach, thought the industry had a "grievance" because of suggestions that favors given by contractors to procurement officers were meant as a "bribe."

"Nearly everyone thinks that taking a DOD employee to lunch improves communication," he insisted. Some of the most important of the NSIA activities are best summarized by two items from the annual report:

WRIGHT-PATTERSON AFB MEETING

Through the courtesy of Gen. Mark E. Bradley, Jr., Commander, AFLC, and Maj. Gen. R. G. Ruegg, Commander, ASD, the Association was privileged to hold its twelfth annual meeting at Wright-Patterson Air Force Base on 22–23 April.

The meeting, classified SECRET, and attended by over two hundred representatives of member companies, included a series of briefings by Gen. Bradley and other distinguished Air Force officers, as well as most interesting addresses by D. W. Smith, President, General Precision, Inc., and Chairman of the Association's Executive Committee; Gerald J. Lynch, Chairman and President of Menasco Manufacturing Company and NSIA's Executive Vice President; and R. E. Beach, Vice President and Corporation Counsel, United Aircraft Corporation, and President of NSIA. Mr. Smith spoke on "Management Evaluation," Mr. Lynch on "The Future of the Subcontractor in the Aerospace Industry," and Mr. Beach on "What Military-Industry Complex?"

The closing event of the meeting was a dinner address on "Air Force Procurement," by Hon. Robert H. Charles, Assistant Secretary of the Air Force (Installations and Logistics).

Meeting at U.S. Army Electronics Command

Approximately 100 representatives of NSIA member companies attended the fifth annual NSIA meeting at the U.S. Army Electronics Command, Fort Monmouth, New Jersey, on 21–22 May. The visit opened with a banquet at Gibbs Hall, where the Commanding General, Maj. Gen. F. W. Moorman, spoke. Briefings on the second day were held in the auditorium of the Hexagon, the headquarters of the U.S. Army Electronics Laboratories, Electronics Command. . . . The final talk of the afternoon was given by Dr. Hans K. Kiegler, Chief Scientist of the Electronics Command, on "The Military-Industry Team—and the Future of Military Electronics."

In 1962 the Department of Defense created the Defense Industrial Advisory Council (the word *Defense* has recently been dropped) as a formal channel for bringing industrial views into the Pentagon. Two thirds of the non-government members are top executives of the fifty leading DOD contractors. Clyde Bothmer, the Executive Secretary of DIAC, describes how it works:

Throughout its wide scope of interests, the council's activities are closely coordinated with those of industry associations also supplying advice and assistance to the Defense Department. Normally, a DIAC sub-group with its DOD chairman supplies advice in an area for which the chairman is responsible within the Department. As policy papers approach a cohesive state, they may be submitted to industry, generally through an appropriate industry association. When the industry comments have been received, the DIAC subgroups may again provide advice as proposed DOD policy statements near final form. Industry associations also assist by nominating members to serve on DIAC supporting groups.

One problem recently taken up by the Council is how to "maintain public and Congressional confidence in the integrity and effectiveness of defense procurement and contractor performance."

A recent report of the Joint Economic Committee of the Congress reveals that 2072 retired military officers of the rank of colonel or Navy captain and higher are now employed by the 100 top defense contractors. This is about three times the number of retired military employed per company ten years ago when the problem first came to the attention of Congressional committees. Major General Nelson M. Lynde, Jr., helped arrange a "sole source" contract with Colt Industries to procure the M-16 rifle. Five months later he was working for the company. Lieutenant General Austin Davis retired from an important role in the Minuteman missile procurement program to a vice-presidency of North American Rockwell, a prime contractor on the same program. These are only two of the many officers who have managed to build two careers on a single weapons system.

An even more significant conflict-of-interest problem concerns the top civilian officials in charge of procurement, research, development, and other Pentagon contracting functions. Recently Senator William Proxmire brought up the case of Thomas Morris, the former assistant secretary of defense for procurement, who became vice-president of Litton Industries. In his last year as procurement chief, Litton contracts jumped from $180 million to $466 million, an increase of 150 percent. Senator Proxmire charged the firm with "buying influence with the Pentagon and plenty of it" through a "payoff." Morris had indeed been "integrally and powerfully involved with every Pentagon decision" on procurement, as Proxmire charged, but his case was by no means unique. A generation of engineer-entrepreneurs, high-level systems managers, and procurement specialists have been shuttling back and forth between defense contractors and strategic positions in the Department of Defense.

This trend began in the early 1950's, with the Korean rearmament and the beginning of the missile program. In 1950 William Burden, a partner of Brown Brothers, Harriman and a director of Lockheed Aircraft Company, was made special assistant for research and development to Air Force Secretary Finletter. A month after he resigned in 1952, one of his partners, James T. Hill, who later was active in Itek (a Rockefeller-financed defense corporation), became assistant secretary for management. Burden was followed in the research-and-development job by Trevor Gardner, president and majority stockholder of Hycon Corporation. Gardner, who had been put in charge of a committee "to eliminate interservice competition in development of guided missiles," resigned in 1956 in protest over the Defense Department's refusal to give the Air Force exclusive control of the missile program. During his years in Washington, his company tripled its government contracts. His successor was Richard Horner, a career scientist who had worked for the military. In less than two years Horner managed to land a senior vice-presidency with the Northrop Aircraft Company. His successsor, Joseph V. Charyk, did even better. He came from Lockheed, and after four years of advising on what the Air Force should buy and build, he became head of Ford's space-technology division and later, after another stint at the Pentagon, president of the Communications Satellite Corporation. His three successors, Brockway McMillan, Courtland D. Perkins, and Alexander Flax, managed to combine similar patriotic service with careers as managers or directors of the Bell Telephone Laboratories, Fairchild Stratosphere Corporation, and Cornell Aeronautics Laboratory, all major defense contractors.

Many principal officials of the Air Force either had a home in the aircraft industry or found one on leaving office. Malcolm MacIntyre, Under Secretary in the Eisenhower Administration, for example, was a Wall Street lawyer who moved up after his Pentagon service to the presidency of Eastern Airlines and later to a vice-presidency of Martin Marietta, another leading contractor. Robert H. Charles, a Johnson appointee who handled the C-5A cargo plane and was prepared to pay Lockheed almost $2 billion more than the contract price, had been executive vice-president of McDonnell Aircraft. Philip B. Taylor, appointed by Eisenhower to be assistant secretary in charge of matériel, had been with both Curtiss-Wright and Pan American. Roger Lewis, another Eisenhower assistant secretary in charge of matériel, who had also been a vice-president of Curtiss-Wright, becamse executive vice-president of Pan American World Airways and finally president of General Dynamics. Another assistant secretary, Joseph Imirie, became yet another vice-president of Litton Industries.

The principal officers in charge of research, development, and procurement for the Army have included William Martin (Bell Labs), Finn Larsen (Minneapolis Honeywell), Willis M. Hawkins, Jr. (Lockheed), and Earl D. Johnson, who, like Frank Pace, a former Secretary of the Army, climbed from the Pentagon into the presidency of General Dynamics. The Navy research-and-development officers, James Wakelin, Robert W. Morse, Albert Pratt, have been associated with such leading defense contractors as Itek, Chesapeake Instruments, Ryan Aeronautics, and Simplex Wire and Cable. From 1941 until 1959 the Navy had an assistant secretary in charge of its own separate air force. Two men who held this post were Wall Street lawyers and directors of aviation corporations: Artemus Gates (Boeing) and John L. Sullivan (Martin Marietta). Two others were connected with the airlines industry: James H. Smith (Pan American) and Dan A. Kimball (Continental Airlines). Kimball later became president of Aerojet General.

Many of the most important decisions on research, development, and procurement have been made at the Defense Department level. Here also the recruitment patterns are the same: in the Eisenhower Administration, Frank Newbury (Westinghouse), Donald Quarles (Western Electric, Sandia Corp.), Clifford Furnas (Curtiss-Wright), to give three examples. In the Kennedy and Johnson Administrations, John Rubel (Lockheed, Litton), Eugene Fubini (IBM, Airborne Instruments Laboratory), Harold Brown and John Foster (Livermore Laboratories).

For the men who sustain the weapons research, development, and procurement process, service in the Pentagon is an essential element of career-building. In virtually every case the individuals in charge of these functions have used their tenure at the Pentagon to better themselves. Many have come as vice-presidents and left as presidents. There is no proof that any of these men has acted improperly, for, indeed, there are few recognized ethical standards to apply. Moreover, no outsider will ever know the real motive for individual procurement decisions. The important point is that their past careers, future expectations, and professional experience make them tolerant of waste in military spending and entirely comfortable with the idea of a permanent arms race. Some will say the Litton Industries–Pentagon shuttle is part of the American tradition. The successful tax lawyer learns the ropes in the Internal Revenue Service. The good broadcasting-industry lawyer may have spent time at the FCC. But the application of this familiar career pattern to the defense business means that virtually the only people making decisions as to whether new weapons systems are needed, whether they cost too much, and who should make them, are men who directly and personally stand to benefit from big defense budgets. In discussing the Morris case noted earlier, Senator Proxmire pointed out that almost 90 percent of defense contracting is by negotiation and not by competitive bid. Thus "whether Litton or some other firm gets a particular contract will be determined very largely by the subjective attitude of Pentagon officials towards Litton officials." Top Pentagon officials exercise especially broad discretion because their actions are subject to grossly inadequate review both in the White House and in Congress. The Bureau of the Budget, which is supposed to review the Pentagon budget for the President, assigns no more than 50 out of 500 employees to the Department of Defense even though the Pentagon uses over 60 percent of budgeted funds. (An index of the Pentagon's power in the federal bureaucracy is the fact that the natural-resources program, which accounts for less than 5 percent of the budget, is assigned 33 employees.)

Thus, America's leaders have built into the weapons-procurement system a set of incentives for continuing the arms race by recruiting the managers of the Pentagon from the arms industry. The taxpayers have been paying for biased judgment. Professor George Kistiakowsky, former Special Assistant to President Eisenhower for Science and Technology, has pointed out that it is not necessary to recruit the managers of the Pentagon procurement programs from the industries that depend upon the defense budget for their survival. The necessary technical knowledge can be found in non-defense industry and in the universities. But profitable patterns developed over a generation are not easily changed. Like other essential links in the military-industrial complex, this one will grow stronger until we recognize that permitting officials to build personal careers by subsidizing defense firms threatens the national security.

The Origins of the Military-Industrial Complex

The government defense-industry nexus defies most of the rules of the free-enterprise economy. The essence of the free-enterprise system is competition, but 57.9 percent of all defense procurement is negotiated with a single contractor and only 11.5 percent through formal advertised competition. Under the capitalist creed, the efficient survive, and those who can neither provide quality nor control their costs fall by the wayside. Defense industry, as we shall see, is shielded by the government from the harsher realities of the competitive system. It is relieved of the obligation to be efficient and is protected by the government from most of the normal risks of doing business for profit. Government and a vast dependent industry have struck a bargain under which industry has surrendered a few management prerogatives to the Pentagon in return for substantial subsidies. We shall look more closely at these subsidies, for they go to the heart of the relationship which we call military socialism. But first there is some essential history to recount. How did the relationship develop?

The military-industrial complex is another legacy of the Second World War. The five-year mobilization transformed the relationship of government and defense industry and gave birth to the symbiotic partnership that exists today. There were few issues on which the two partners entered the wartime association thinking alike. Before the war most businessmen looked upon the military as boondogglers and practitioners of an obsolete and barbaric art. Some of the most prominent businessmen, such as Andrew Carnegie and Henry Ford, were commercial pacifists who believed that the salvation of mankind lay in trade expansion. One day the whole world would become One Great Market. The choice for America, as they saw it, was between industrial progress and militarism, and the latter they equated with waste. The professional military were equally contemptuous of businessmen, who, in their view, constituted a money-grubbing leisure class. Businessmen "never consider the propriety of devoting themselves or their sons to the public service unless it be as ambassadors or ministers at foreign courts," General William Carter complained in a 1906 number of the *North American Review*. Patriotism was one thing and profits were quite another. There were, to be sure, a few exceptions. The Navy League, a lobby composed of naval officers and shipbuilders devoted to extracting more money from Congress for ships, was an early prototype of the military-industrial complex.

Big business was dragged reluctantly into rearmament in 1940 by the New Deal. The War Resources Board, under Edward Stettinius, president of U.S. Steel, set modest rearmament goals despite the fact that the steel industry was operating at less than 54 percent capacity at the time. The New Dealers with the biggest reputations for being anti-business, Henry Wallace, Harry Hopkins, and Leon Henderson, were constantly pressing for faster and more drastic rearmament than the businessmen in charge of procurement and the Army itself. Generals were aghast when Roosevelt set a production goal of 50,000 planes a year.

Business feared full-scale mobilization because they were nervous about being saddled with excess productive capacity if the war should turn out to be too small. The model on which they operated was World War I, in which the government manufactured most of its own armaments in federal arsenals. General Motors kept on producing automobiles all during the First World War and did a total of $35 million in war work without expanding its plant at all. In World War II, on the other hand, G.M. had $12 billion in government contracts. From February 1942 to September 1945, the company did not produce a single automobile. At the same time it went into production on 377 totally new items. Other firms were even more totally transformed by the war. Whole new fields, such as radar, advanced communications, and nuclear energy, were opened. Entire new industries which had not existed in the pre-war period were now flourishing captive suppliers of the federal government.

The relationship that evolved in the war was a marriage of necessity, but the government brought to it a handsome dowry. Contracts went to the companies that were able to marshal talent and facilities quickly. These were the biggest corporations in the nation. Forty percent of all research contracts, for example, went to the ten top corporations. The government financed hundreds of new plants and then at the end of the war sold many of them to the industrial giants at nominal cost. The Geneva Steel Plant, which cost $202.4 million to build, was given to U.S. Steel for $47.5 million. The government also paid for research that ended up as private patents. It gave fast tax write-offs, and at the end of the war terminated contracts on terms so favorable to the contractors that the Comptroller-General later found that in one out of every seven cases the contractor received excessive or fraudulent payment.

Big business had faced a crisis in the war. Some, like Bernard Baruch, proposed that the government nationalize the war effort, draft industry into the war economy, and set industry profits itself. General Brehon Somervell, chief of the military procurement agency, the Services of Supply, with 42 percent of the annual war budget to spend, called for the total militarization of the economy. The Army would take over the plants. Corporation presidents would wear uniforms and be subject to military law. But the mobilization crisis was solved by carrots instead of sticks. The guiding principle of the war effort was to give business a variety of incentives to produce. The free-enterprise system fueled by government funds caused the Gross National Product to more than double in five years. A galloping economy—$687 billion in Gross National Product in the war years—overwhelmed the enemy with America's productive might. But in the process the system was transformed. In 1944 Charles E. Wilson, president of General Electric, speaking to the Army Ordnance Association, proposed a "permanent war economy." What he had in mind was a permanent set of relationships between business and the military which could be the nucleus of any future general mobilization and the conduit for the substantial military production he assumed would continue in the postwar world. Every major producer of war materials, he advised, should appoint a senior executive with a reserve rank of colonel, to act as liaison with the Pentagon. "There must be once and for all," he said, "a continuing program."

By the fall of 1945, two months after the end of the war, both the Air Force and the aviation industry agreed that national security was still a big and expensive job. General "Hap" Arnold, Chief of Staff of the Army Air Corps, had warned in early 1945, while the war was still on, that the U.S. would be the "first target" of the next war. On Armistice Day, 1945, he told the public that to guard against surprise attack, the U.S. "must be prepared to strike back with 3000-miles-per-hour robot atom bombs launched from space ships." The following month *Aviation* Magazine featured an editorial on "Reconversion to Recomplacency." Conceding that "at this moment in history we are able to conquer any nation or combination of nations," it warned that national security requires us to "set up a demobilization plan based on the assumption that we *shall* be attacked without warning."

The experience of war made American business

ready to accept the military as a permanent, legitimate force in American life with which industry could form profitable coalitions. By the same token, the war rehabilitated the public image of big business, which had been badly tarnished in the depression years of the 1930's. In creating the Arsenal of Democracy, business demonstrated its capacity for patriotic service. There was no doubt that what was good for General Motors was good for the country when G.M. was producing tanks to fight Hitler. In addition, the rapprochement of business and the military was eased by the changing character of the military task. As Professor Morris Janowitz has noted, "the civilian character of the military establishment increases as large numbers of its personnel are devoted to logistical tasks, which have their parallels in civilian enterprise." As the military became increasingly involved in problems of technology and politics, and particularly with the job of managing billions of dollars' worth of assets in the civilian economy such as factories, hospitals, housing, and transportation, the historic distinctions between the soldier's heroic ethic and the capitalist creed faded.

Military Socialism

It is hardly surprising that leading defense contractors like General Dynamics keep insisting that "we're in [defense] business to stay." According to *The Economics of Military Procurement,* a report of the Joint Economic Committee of the Congress, known as the Proxmire Committee Report, these companies receive a "vast subsidy" in tax dollars. Military socialism offers unique economic advantages to those firms that have

come to take the place of the government-owned arsenal. They exercise broad public power and make substantial private profit. How does the system work?

As already mentioned, the defense contractor devel-
ops his new products and the market for them simultaneously, often in close, continuous association with the
customer, the Department of Defense. Thus he can
normally count on selling them. The government does
reserve the right to make certain decisions about the
internal operations of the firms which in the private
economy are exclusively the prerogative of management. For example, the Pentagon insists upon the right
to pass on subcontractors, to decide which products
should be purchased in the U.S. rather than abroad,
and what minimum as well as average wage rates will
be paid. The government insists upon rights of cancellation of contracts which private purchasers do not have.
The fact that this power is occasionally exercised,
sometimes with serious disruptive effects, may account
for the recent unimpressive showing of defense issues in
the stock market. Most defense stocks are down from
their all-time highs, in part due to recent concern about
the levels of military spending, and in part to the fact
that defense stocks pay low dividends. But for the principal defense contractors themselves, doing business
with the Department of Defense continues to offer
unique privileges in the American economy.

Such privileges account for much of the waste in
defense contracting. The Proxmire Committee Report
gives a picture of the dimensions of such waste:

> In the past, literally billions of dollars have been
> wasted on weapons systems that have had to be
> canceled because they did not work. Other systems

have performed far below contract specifications. For example, one study referred to in the hearings shows that of a sample of 13 major Air Force and Navy aircraft and missile programs initiated since 1955 at a total cost of $40 billion, less than 40 percent produced systems with acceptable electronic performance. Two of the programs were canceled after total program costs of $2 billion were paid. Two programs costing $10 billion were phased out after 3 years for low reliability. Five programs costing $13 billion give poor performance; that is, their electronics reliability is less than 75 percent of initial specifications.

Why does the Pentagon continue to distribute its immense supply of favors to a selected few? Do the top ten or twenty contractors "control" the Pentagon? Each of the companies which ranked among the top ten defense contractors over the seven years 1961–1967 was still among the top fifteen in 1968. Yet individual fortunes do rise and fall. Thiokol makes 96 percent of its sales to the government. In 1961 it was awarded $210 million in military contracts and was the 28th-ranking defense contractor, but by 1968 Thiokol was down to $119 million and 58th place. Boeing in 1962 had contracts with the Department of Defense totaling more than a billion dollars; in 1968 the military side of the corporation was down to $762 million. The second-ranking contractor, Lockheed, did not have the power to prevent the Pentagon from canceling the multimillion-dollar Cheyenne helicopter contract when it came to public attention that the rotor tended to fly off and the plane was likely to cost far more than the contract price.

The ultimate power to distribute the $45 billion now spent annually on procurement resides in the Pentagon itself. No other unit in government or the private economy comes close to the Office of the Secretary of Defense in money to spend, power to decide how to spend it, or respectful attention from every other center of power in the society. The defense contract is much like the royal franchise under which the King used to give private interests the right to make money from colonies while at the same time advancing the interests of the Crown. Much of the power of the Pentagon, like that of the King, derives from its broad discretion to make private fortunes for its clients. But, as Columbia Professor Seymour Melman has noted, it is the Pentagon and not the clients whose interests are paramount. In the Department of Defense the leading defense contractors are regarded as subsidiaries that must be protected for the benefit of the whole system.

A defense contractor is very little more than an organization of managerial and technical capabilities. The government owns everything else. For example, LTV Aerospace Corporation owns 1 percent of the 6.7 million square feet of office, plant, and laboratory it uses. The rest it leases from the Department of Defense. To military planners, however, "the team" of managers and engineers is a national asset. The Pentagon is prepared to be lenient with organizations that can innovate in unknown and untried systems. Weapons laboratories that function on the "frontier of technology" are not allowed to fail, for "the team" must stay together and keep the laboratories "hot." If LTV Aerospace did not exist, it would have to be invented. The evidence is persuasive that the Secretary of Defense awarded the TFX plane contract to General Dynamics, against the unanimous recommendation of the Joint Chiefs of Staff that it be given to Boeing, in order to rescue a ranking contractor from probable collapse. General Dynamics, which had lost over $400 million on its Convair division a few years earlier, needed a substantial military contract to survive.

Since there is no catalogue price for a missile and no shortage of money in the Pentagon, procurement officers have had little reason to maintain tight controls over contractors. A military officer whose career rises or falls with a weapons system finds it easy to justify waste in procurement as insurance against technological scarcity.

Under McNamara the Department of Defense increased its power over the economy in a number of respects. William D. Phelan, Jr., has analyzed some of the specific consequences of the McNamara revolution in a report for the Ripon *Forum*. He points out that McNamara, whom he calls "the business schools' idea of God," succeeded in concentrating power in the Office of the Secretary of Defense by the use of advanced management techniques backed by an inexhaustible supply of computers. This gave the management of the Pentagon enormous advantages in dealing with the other agencies of government and private contractors too. Pentagon power was further enhanced by the establishment of a number of new agencies within the Office of the Secretary of Defense. Not only has the Defense Supply Agency centralized authority over purchasing for the military, but in such areas as petroleum, for example, it is now assuming the same power for all government agencies. The Defense Documentation Center stores technical information that it makes available to eligible contractors. As Phelan notes, the power to grant or withhold access to the nation's most valuable collection of technical data is "a powerful economic weapon." The Defense Commercial Communications Office has given franchises to more than 300 commercial communications carriers and encourages the nonmilitary federal agencies to go through the Pentagon in fulfilling their telecommunications needs. This increased dependence of non-military agencies on the incomparably superior technical and managerial facilities of the Department of Defense is, according to Phelan's study, a general trend. Giving "hegemonic power to the Pentagon," he declares, "constitutes nothing less than a tendency towards the militarization of the entire government." One index of the Pentagon's power is that it owns substantially more than $200 billion in assets, including real-estate holdings about the size of the state of New York.

In addition to the power over procurement, technology, communications, and fiscal management already discussed, the Pentagon is now involved in social engineering as well. Under McNamara and Clifford, the Pentagon trained hard-core unemployed. Secretary Laird has a plan to get kids off ghetto streets by sending them to "camp" at military installations. The Pentagon has established its own analogue to the White House Urban Affairs Council to coordinate its increasing role in the management of America's social problems.

The concentration of decision-making in the Office of the Secretary of Defense has given the Pentagon a power which no combination of defense firms can match. The Department of Defense has become the closest thing to a central planning agency in American society. It uses the military budget to stimulate economic growth, to put money into circulation in times of recession, to encourage the development of specific industries, and to assist certain geographical areas.

The Politics of Defense

The alliance of defense contractors and politicians is another cornerstone of the military-industrial complex. Six months after his appointment to the House Naval Affairs Committee in 1937, freshman Congressman Lyndon B. Johnson obtained a major defense contract for his principal financial backers, the Brown & Root construction firm. The same firm thirty years later was called upon to turn South Vietnam into a succession of military bases at considerable profit. In the Johnson years Texas moved ahead to become the third-ranking state in military contracts. Between 1962 and 1967 the value of prime contracts awarded to Texas firms increased by 350 percent.

William D. Phelan, Jr., has studied the political ties of five of the defense companies that experienced the most dynamic economic growth in the 1960's. Each of the following companies increased its revenues by more than 500 percent: Litton Industries, Ling-Temco-Vought, Gulf & Western, Teledyne, and McDonnell Douglas. Phelan suggests that the following information about these companies is not wholly irrelevant:

The chairman and chief executive of Litton is Charles B. Thornton, a member of the Defense Industrial Advisory Council, long-time associate of McNamara, and close friend of President Johnson. The top man at L-T-V is James J. Ling, one of Humphrey's leading supporters, a business ally of several of Johnson's long-time Dallas backers, and the holder of corporate control over several subsidiaries in whose management Abe Fortas and his law partners have been particularly active. Perhaps the most powerful outside director at Gulf & Western is Edwin L. Weisl, Johnson's most loyal backer in New York. Cyrus Vance went to the Defense Department from Weisl's law firm. The top man at Teledyne, Henry G. Singleton, has long associations with both Thornton and Howard Hughes, an influential man in Johnson's background during much of his career. Teledyne's co-founder and a powerful director is George Kosmetsky, the dean of the business school at the University of Texas.

Finally, McDonnell Douglas is well-connected to the Missouri branch of the Democratic Party. In addition to ties to Clark Clifford and Stuart Symington, it at one time included among its directors James E. Webb, until recently head of NASA, and a close advisor of both Johnson and the late and very powerful Senator Kerr of Oklahoma.

When one scores five out of five, desultory muckraking becomes structural revelation. As we have seen, all the biggest corporate winners have had powerful political contacts at high levels in the government and in the Democratic Party. This does not signify corruption so much as the realities of the military-industrial system.

In the 1960 Presidential race the Democrats, using erroneous intelligence estimates leaked by the Air Force, cried "missile gap" and distributed hundreds of thousands of copies of a booklet calling for "an impressive additional expenditure of about $4 billion a year on our strategic forces." John F. Kennedy personally played the preparedness theme to the hilt: "I am convinced that every American who can be fully informed as to the facts today would agree to an additional investment in our national security now rather than risk his survival, and his children's survival, in the years ahead." The Kennedy advisers, who had shrewdly estimated that the number of voters opposed to survival was small, came ten years later to see that such demagogic appeals undermined national security. In 1968 Richard Nixon, speaking a few hundred yards from a General Dynamics plant, charged the Democrats with a "security gap." This time the public was more skeptical, mostly because of their disillusionment with the Vietnam war. Nevertheless, in the same election at least one U.S. Senator, Joseph Clark of Pennsylvania, was helped to defeat by the efforts of a union upset about his interest in cutting the defense budget and closing down defense installations.

Most Congressmen and Senators, who have little personal knowledge of defense matters, have generally concluded that the political climate in the country demanded continued support for high-level defense expenditures. Many of them have found it convenient to run against Khrushchev or Mao on a "preparedness" platform. Once elected, the safe and easy course was to leave the problem in the experienced hands of the chairmen of the Armed Services and Appropriations Committees—four conservative Southern politicians, Senators Stennis and Russell of Mississippi and Georgia, and Congressmen Rivers and Mahon of South Carolina and Texas. The most famous example is L. Mendel Rivers, House Armed Services Committee Chairman, whose campaign slogan "Rivers Delivers" magnificently understates the tribute which the Defense Department has showered on his hometown, Charleston, South Carolina.

However, the Congressional wing of the military-industrial complex has been careful not to hoard everything. "My friends, there is something in this bill for every member," Chairman Carl Vinson of the House Armed Services Committee exclaimed in 1958 as he presented a billion-dollar military construction bill to the whole House. Defense contracts are currently distributed among 363 Congressional districts. Congressmen have the incentive to vote right on military appropriations. Many are cultivated with junkets, testimonials, and other diversions.

Until late in the Vietnam war, the budget of the Department of Defense was traditionally accepted unanimously in both houses after a few hours of patriotic speeches. "I never voted against a defense appropriation," former Congressman and later Secretary of the Interior Stewart Udall testified before the Joint Economic Committee. Such appropriations, he said, were "sacrosanct." The Armed Services Committees and the Appropriations Committees are the only Congressional bodies which make any pretense at reviewing the defense budget proposed by the Administration. "I've got the most powerful position in the U.S. Congress," Mendel Rivers once boasted to a colleague. As Chairman of the House Armed Services Committee and a close friend of Speaker McCormack ("Where the national interest of our country is concerned, I would rather err on the side of strength. . . . I am willing to follow the leadership of the gentleman from South Carolina"), Rivers rules the Committee with an iron hand. He and the chief staff counsel, John R. Blandford, a brigadier general in the Marine Corps Reserve, constantly propose additions to the budget beyond what the Department of Defense requests. In 1968, for example, Rivers proposed making a $14 million payment on a $62 million subsidy to Northrop Aviation to enable the company to redesign the F-5 "Freedom Fighter" to compete with the French Mystère for export sales. Even though other members of the Committee wondered under what theory the taxpayer was making this gift, Rivers' amendment passed the House 341 to 21. Members of the Committee who find themselves in disagreement with the Chairman also find that their access to information is severely limited. Each member now has five minutes to question witnesses. Only Pentagon witnesses and occasional outside consultants to the Department of Defense are invited to testify. Rivers' favorite question is: "Are you sure you have enough?" Typical of the Chairman's scrutiny is the following item from recent hearings:

> MR. RIVERS: Let's get to missiles. Are there any questions on airplanes? We need them so bad I can't conceive how there can be. Let's get to the missile program.

Warren Unna of the Washington *Post* interviewed several members on the Committee's day-to-day operations. He shows how the group which is supposed to function as a watchdog over the military actually functions:

> Members out of favor with Rivers say they find themselves showing up at the closed-door 10 a.m.

meetings expected to raise their hands and vote "aye" to a pile of documents they've had no time to study. Hearings usually complete their cycle on Thursday afternoon. Minority views must be in by Friday. The military authorization then quickly moves to the House Rules Committee the following Monday. Even committee members have their first chance to see the actual committee report when the bill is on the floor—and usually passed —on Tuesday.

III

Choosing Life:
A Strategy

The central task of American society is to free ourselves from the Economy of Death. Whether we shall have the money, energy, or will to turn back the rising tide of violence at home, whether we can rouse ourselves from the dream of Pax Americana abroad, are equally dependent on the same critical question: Are the American people capable of reordering national priorities so that their children can live as free men and in peace?

In politics, as in medicine, the first step toward health is to diagnose the sickness. You do not undergo surgery for the common cold, or treat cancer with a steam kettle. Unfortunately, the cancer analogy is uncomfortably accurate. American militarism is a systemic disorder that has affected each of our major institutions and distorted their development just as cancer distorts the development of the human cell. We can hope the cancer is still operable, but time is short.

Amid the recent concern over the militarization of America, two types of critics have emerged. There are the optimists who talk about controlling the military-industrial complex and the pessimists who talk about breaking it up. J. K. Galbraith, one of the optimists, believes that "the problem of military power is not unique." The military-industrial complex represents merely one, albeit a formidable one, of a number of bureaucratic institutions that need controlling. Happily, the Pentagon has sown the seeds of its own destruction by its cavalier attitude toward waste, its mismanaged wars, and its weakness for the patriotic lie. Aware that the communist world is no longer "monolithic," deeply concerned with domestic insecurity, and disillusioned with counterrevolutionary crusades, the public has begun to regard the military with skepticism. The military is not, Galbraith concedes, "on its last legs," but it is on the run. He believes that an "ecumenical" campaign to unseat the reactionary Southern Senators and Congressmen who run the crucial Congressional committees, to obtain independent scientific experts to advise Congress, and to elect a sympathetic President is about all it will take to set national priorities aright.

It is true that the day of the blank check for the military is over. There is more questioning about military waste and distorted priorities than at any time since World War II. Nevertheless, Galbraith's self-styled "moderate" program for controlling the military is a poultice and not a cure. Military spending is indeed out of control, but control as an end in itself is not the answer. As long as the military-industrial complex continues to play anything like its present role in American society, it will remain uncontrollable. Congressional opposition can complicate the life of the Pentagon and perhaps slow its pace, but Congress cannot bring about the fundamental reordering of priorities needed for national survival if it sees its role as merely policing the Pentagon as we know it. Nor can a budget-conscious President do much more unless he has a strategy for political reconstruction which makes it possible to reduce radically the power of the Pentagon in American life.

Military spending is not primarily an auditing problem. It is a problem of the illegitimate exercise of political power. It might be possible to make the Pentagon more efficient by better scrutiny of extravagance, profiteering, and graft; but that would merely enhance its power. Cancellation of one or two particular programs such as the ABM, or even shaving $5 billion to $10 billion from the budget as a whole, is not enough. There are now waiting in the Pentagon wings thirteen major new weapons systems which will move inexorably toward production and deployment unless the juggernaut is stopped. In the past when the Pentagon has run into Congressional opposition on a weapons system, it has halted for a moment, regrouped, and mounted another assault. As Richard Kaufman of the Proxmire Subcommittee has pointed out, old Pentagon systems never die; they hibernate. The ABM was originally designed to meet a nonexistent threat (a developed Soviet ABM), and when the absence of this particular threat was noted, another series of threats was conveniently discovered. MIRV has a similar history. The B-70 manned bomber was finally felled after a mighty McNamara struggle, but it has been back in the budget for several years now with a new set of initials (AMSA) and more recently has been dubbed B-IA. A Congressional audit group would have its hands full merely keeping track of the aliases which the services give to weapons systems that run into trouble.

The pessimists believe that the concentration and entrenchment of power in the military-industrial complex must be broken before it will become possible to control military spending. The sources of energy for the military juggernaut are the interrelated institutions discussed in Chapter II, each of which, by playing its appointed role, pushes the society toward greater militarization. As long as these institutions can command the resources, talent, and public approval which they have taken for granted for so many years, they will be able to defy or frustrate all efforts to make them use their money and power well. Thus, more effective control of the Pentagon is essential to a strategy of conversion, but it is only a technique. The more fundamental task is to shrink the Pentagon to a manageable size so that it can be made properly subordinate to the civilian government, and to alter the institutional relationships that support the Economy of Death. The military-industrial complex is a collection of powerful incentives toward death. If we are to change direction, we must now build into the economy powerful incentives toward life.

Breaking up the military-industrial complex does not mean unilateral disarmament, as President Nixon has charged, except in the sense that any act of self-restraint can be so characterized. It does not mean closing down the Pentagon or leaving the country any more naked to attack than it is already. It does mean drastically downgrading the role of military power in foreign policy and increasing security through arms control, the peaceful solution of international differences, and the development of a new world environment that can better support human life.

Reorienting the way the United States relates to the outside world requires a major reconstruction of those domestic political and economic institutions that are pathologically dependent upon military spending. Militarism has had thirty years to spread through American society. It cannot be arrested overnight. As we have seen, there are powerful forces in America with strong emotional, economic, and political ties to the Economy of Death. The only force capable of bringing about the conversion of the society is a nation-wide movement of Americans who see the militarization of America as our number-one national-security problem and are prepared to fight for a generation if necessary to free the nation from its grip.

Any movement dedicated to rescuing the American people from the Economy of Death has two principal tasks. First, it must persuade enough Americans that neither peace nor security is to be achieved through the endless preparation for war. Second, it must hold up an alternative vision of an American economy that can function without military socialism and an American society that does not feel compelled to seek security through violence.

This chapter sketches a strategy of change. The first part describes a program of specific steps that might be taken by the Administration and the Congress to restore balance in the national priorities and to move the nation away from militarism. The second part discusses what the citizen can do to create the necessary political climate to make this happen.

Improving the Machinery for Setting
National Priorities

"What, quite frankly, we didn't look at was whether we actually needed a defense system of this magnitude at all," Charles L. Schultze, Director of the Bureau of the Budget in the Johnson Administration, recently testified before a Congressional committee. "At the time nobody was asking these questions."

One reason that nobody asked them was that there has been neither a tradition of asking them nor an opportunity to do so. To a great extent the Pentagon itself decides what it needs. The "wish list" of the Joint Chiefs of Staff, always deliberately inflated for bargaining purposes, is whittled down by the Secretary of Defense, who then negotiates with the Bureau of the Budget. For all other agencies of government, according to former Budget Bureau Director Schultze, the Bureau of the Budget makes the final recommendations to the President. For the Pentagon, however, the Secretary of Defense has made his recommendations directly to the President, and the Budget Director's role has been limited to making comments. The President has made the final decision on the basis of his best political judgment. Nowhere in this process is there machinery for adequate review of the Pentagon's budget. Most important, there is no context in which to balance domestic needs and military requirements.

To decide whether a new piece of military hardware is needed, you need to know the answers to three questions: First, what will you be able to do with this weapon that you can't do with what you have? Second, is this a capability we should want? Third, can we afford it? The first question strikes at the problem of waste and duplication. The second goes to the far more crucial issue of the political justification of new military systems—i.e., whether they make the world more dangerous or less dangerous. The third question can be answered only by comparing rival claims on the tax dollar. Whether the U.S. can afford a new aircraft carrier or extra missiles depends upon whether we need breathable air more.

There is at present no forum in the Executive for planning national priorities in a rational way. The Cabinet no longer deliberates as a body. The National Security Council includes no one to argue for a distribution of resources other than what the Pentagon demands. Except for informal ad hoc discussion among the President and those advisers he happens to consult, there is no opportunity to argue for an order of priorities different from the one the Pentagon sets.

The National Security Council should be reconstituted to reflect a broader definition of security. The Secretaries of Transportation, of Housing and Urban Development, of Health, Education, and Welfare, and other spokesmen for domestic interests should be members. Such a forum could engage in continuous debate and review of the three competing demands for the tax dollar: (1) public funds for defense against foreign military threats; (2) public funds for defense against such domestic threats as poverty, disease, ignorance, and the poisoning of the environment; (3) private spending money for the consumer-taxpayer. The Pentagon would have to justify its claim to $11 billion for an ABM, say, over rival claims for 400,000 units of low-cost housing, 400,000 additional schoolrooms, and 1300 hospitals, all of which could be bought for about two thirds of the price. Only in such a context is it possible to determine what is a proper tax rate. Taxpayers, who understandably resent present high tax rates because over 70 cents of the tax dollar goes for unproductive military or para-military expenditures, might be willing to accept even higher taxes if they received value for the money they give the government. There is no way to achieve mathematical precision in comparing defense, non-defense, and private-consumption needs. It is possible, however, to provide a framework for making an informed political choice whether it is safer to say no to the Pentagon for additions to the nuclear stockpile or no to the Mayor of New York when he asks for funds to save a collapsing city. The exercise of such political judgment can mean the difference between survival and catastrophe.

The Congress should greatly strengthen its own capacity to reassert control of military spending. The Report of the Congressional Conference on the Military Budget and National Priorities released in June 1969, which was endorsed by thirty-five Congressmen and nine Senators, made two proposals to do this. First, the Report recommended the establishment of a Defense Review Office which would provide Congress with its own expert advice to enable individual Congressmen to evaluate military programs and to scrutinize the Pentagon budget. With such assistance, Congressmen would be able to make a judgment as to whether a new airplane such as the F-14, which will cost at least $20 billion, is a justifiable expenditure, or whether bases acquired more than twenty years ago should be retained. The Defense Review Office would provide the necessary information and insights to bring some of the basic assumptions behind the defense budget to public debate. Congressmen could ask the same kind of searching questions on defense spending they like to ask with respect to the poverty program. With adequate access to information, the Congress could stop acting like a rubber stamp on military appropriations. Second, the Report suggested the creation of a Congressional Joint Committee on National Priorities. This committee would provide a continuing forum for evaluating defense requirements in the context of other needs. It would play the same role for the Congress that the expanded National Security Council recommended above would play for the Executive.

The seniority system is usually blamed for the fact that unbeatable old Southern Congressmen from one-party districts dominate the Armed Services Commit-tees. Obviously, the seniority system frustrates demo-cratic government and should go. But the reform of the defense process cannot wait that long. Membership on the Armed Services Committees should be subject to special rules, in view of the supreme importance of these bodies. There should be rotation of membership on these committees so that every Senator and Con-gressman who serves on them has also had significant experience with competing domestic needs. No chair-man should ever again be allowed to serve more than six years. The country cannot afford the luxury of having members of the legislative branch acquire the obligations to the military or wield the power which Senator Russell and Congressman Rivers have amassed during their reign.

In the final analysis, however, Congress can play its Constitutional role of restraining the Executive only if individual members of both houses are prepared to vote against military appropriations which in their view are foolish, dangerous, wasteful, provocative, or immoral. As long as the Pentagon can count on ultimately ob-taining the funds, it will be prepared to ride out debates and hostile speeches. Only when significant numbers of Congressmen are determined to vote against all or part of the defense budget and to speak out at length in explanation of their vote will Congress recover the power over the military which has slipped away over the last generation.

*How to Review and Control Foreign-Policy
Commitments*

Apologists for big defense budgets have a routine
answer to critics who want to cut military spending: the
budget depends upon "commitments" and "responsi-
bilities," and since these are immutable, nothing can be
done to reduce defense spending. This argument can be
challenged on two grounds. First, it is by no means
clear that there is any logical connection between many
of our commitments and much of the hardware we buy
to maintain them. For example, how will the proposed
new carriers advance our interests better than the
old ones? Second, as we saw in Chapter I, a number of
the assumptions used to justify "commitments" are
highly questionable. For example, the U.S. spends $50
billion a year on its "General Purpose Forces," which
are used in a world-wide struggle against history that is
bound to cost more than any nation can afford.

The Secretary of Defense issues a Posture Statement
each year which explains why the Department of Defense wants so much money and what it plans to buy
with it. The statement neither identifies adequately the
cost of particular forces nor gives an accurate picture of
what current procurement decisions will cost in future
years. Congressmen are expected to authorize new
weapons and forces without knowing how much money
these will ultimately require. Nor does the present Posture Statement attempt to show why particular pieces of
hardware are needed to meet particular foreign-policy
commitments. The Secretary should be required under
law to justify each request for additional funds to mod-
ernize the forces or to add to them by demonstrating
the essential political function they will serve. The bur-

den of proof should be on the Secretary to justify the additional expense by showing why the new weapons are needed to meet specific commitments.

The fact that the Secretary of Defense rather than the Secretary of State issues the annual Posture Statement is a symbol of the militarization of our foreign policy. Charles Schultze has suggested that the Secretary of State be required to submit an annual Posture Statement to the Congress. "This statement," he writes, "should, at a minimum, outline the overseas commitments of the United States, review their contribution or lack of contribution to the nation's vital interests, indicate how these commitments are being affected and are likely to be affected by developments in the international situation, and relate these commitments and interests to the military posture of the United States." The burden of proof should be on the Secretary to justify the maintenance of each overseas commitment. Commitments should no longer be considered unreviewable merely because they were made a long time ago.

The Symington Subcommittee of the Senate Foreign Relations Committee, which has been conducting a long-overdue independent review of overseas commitments, should be made a permanent body of the Congress. It should conduct intensive annual hearings with the Secretary of State and other foreign-policy officials to scrutinize the official justifications for maintaining overseas commitments. The Subcommittee might make recommendations as to which commitments should be dropped. When appropriate, it should draft amendments to the defense-appropriation bill directing that no further funds be spent on them.

These and other hearings of major educational value for the public should be televised on a regular basis. At present the press ignores the educational hearings of the Senate Foreign Relations Committee unless they are staged as confrontations with the Administration. "The Foreign Relations Hour," a TV program in prime time in which committee members of both parties would raise critical national-security issues and answer questions from the public, should become a bi-weekly event.

Congressmen and Senators should also hold local hearings in their own districts at which constituents would have the opportunity to raise questions and give comments on matters related to national priorities, the defense budget, and overseas commitments. Congressmen should invite spokesmen for the Administration as well as critics of Administration policy to appear. Such hearings would contribute to an informed debate on these crucial issues and would force Congressmen to commit themselves on defense and national-security matters so that the electorate could pass on and attempt to influence their views.

All of these proposals are designed to encourage an informed public opinion which can back up Congress' efforts to reassert its Constitutional role in foreign relations. Secrecy and cynical manipulation of the public and its elected representatives, according to Merlo Pusey's recent study *The Way We Go to War,* have been the standard practice of Presidents for a long time. Congressman Abraham Lincoln, outraged by President James Polk's unconstitutional undeclared war against Mexico, analyzed the danger to America in letting the President use the power to commit our military forces abroad:

Allow the President to invade a neighboring nation, whenever he shall deem it necessary to repel an invasion, and you allow him to do so, whenever he may choose to say he deems it necessary for such purpose—and you allow him to make war at pleasure. Study to see if you can fix any limit to his power in this respect, after you have given him so much as you propose. . . .

The provision of the Constitution giving the warmaking power to Congress, was dictated, as I understand it, by the following reasons: Kings had always been involving and impoverishing their people in wars, pretending generally, if not always, that the good of the people was the object. This, our Convention understood to be the most oppressive of all kingly oppressions; and they resolved to so frame the Constitution that no one man should hold the power of bringing this oppression upon us.

Every President in the postwar period has insisted upon the right to enter into new foreign commitments of a military nature without Congressional or public consent or even knowledge. Thus, General David A. Burchinal negotiated an agreement with the Spanish government for a renewal of U.S. bases in Spain which committed the U.S. to aid Franco in the event of a domestic uprising against his Fascist regime. The Chairman of the Senate Foreign Relations Committee would have known nothing about it had a newspaper reporter not happened to discover it. Decisions that are shielded from review run special risks of being wrong. The secret involvement in Vietnam led to a national disaster, and the secret involvement in Thailand now threatens to turn into a second Vietnam. Once a secret commitment is made, extrication is difficult. Thousands must die to rescue the national honor.

Since military spending is related to foreign commitments, it cannot be controlled unless the process by which foreign commitments are made is also subject to control. Such hasty interventions as the 1958 Lebanon invasion or the 1964 Congo operation, which President Nixon has cited as proof that the President needs unlimited power to send American boys overseas, turn out to have been ill-considered and unwise. Except for an immediate retaliation against a nuclear attack on the U.S., there is no reason to grant the President the power to undertake swift military action abroad and there are a good many reasons to withhold it. The Pulitzer Prize journalist Merlo Pusey has proposed a War Powers Act which would prevent the President from stationing troops abroad except in connection with a treaty ratified by the Senate. The President should also be required to submit regularly a complete public list of U.S. commitments and military installations abroad. The emergency powers of the President to send U.S. troops abroad to protect U.S. lives and property should be subject to strict review, since it has consistently been used as a pretext for counterrevolutionary action against other countries. Any foreign military operation should be subject to automatic Senate review.

How to Break Up the Military-Industrial Complex

In 1963 *The New York Times* conducted a survey of the top twenty-five defense contractors and found that none of them had done any serious planning for conversion to non-defense production. Five years later Bernard Nossiter found in a similar survey that the heads of defense firms, some of whom in the interim had experimented unsuccessfully with non-military technology, had no intention of converting their plants to peace-oriented production. The major defense corporations have given no support to the efforts of Senator George McGovern and a few others in Congress who have tried to develop a federal program to aid the conversion process. The presidents of the five largest defense contractors refused even to appear before Senator Proxmire's Subcommittee on Economy in Government to discuss how to end waste in the procurement process. Contractors derive too many advantages from military socialism to give it up easily.

Where in the civilian market can a contractor obtain a public franchise of equal value? His prices need not be competitive because there is usually no competition. The Pentagon is the only customer, and it is tolerant of weapons that do not meet specifications or that cost several times what they were supposed to cost.

Several contractors who have tried to develop civilian products have found that they cannot compete in the non-military market. Their technology is often too fancy and too expensive for customers who do not have billions to spend. "Gold-plating" is a common practice. The military services are glad to pay for unnecessarily sophisticated or complicated hardware since money is plentiful and promotions go to enterprising officers who know how to get money and prestige for their services by pushing technology to the limit. In the civilian economy, on the other hand, customers will not pay for extras. Defense contractors presently have no incentive to convert to the Economy of Life, and they do not believe that they will ever have to do so. As long as the U.S. shares the planet with several billion neighbors, the supply of threats is inexhaustible. There is a weapons "revolution" every five years, a new piece of lethal technology which can always be sold. Despite recent attacks on the military, the weapons-makers do not see the political forces in America which will force a shift in national policy sufficient to require them to reorient their corporations. Until they do, they will take no initiative toward conversion, but will use their power to delay it.

Thus the entrenchment and concentration of economic and political power in the hands of the military-industrial complex is itself a prime obstacle to conversion. The public-relations and lobbying activities of the Pentagon play an important role in building and protecting that power. There is no legitimate reason for the armed forces to support over 6000 public-relations men. If the Defense Department has a significant job to do, it should not have to advertise. It is not the job of the military to educate the country or to provide thrills for the population. Congress should drastically cut the public-relations budget and prohibit the Pentagon from propagandizing the American people.

The taxpayer should not have to pay for 339 Pentagon lobbyists with $4.1 million a year to spend on pushing the service viewpoints with Congressmen and Senators. "Legislative liaison" functions should be drastically curtailed. Technical military information should be provided to the Congress by their own enlarged staffs, the proposed Defense Review Office, and independent outside consultants. The Pentagon can present its story just as other departments do, through legislative hearings and modest liaison efforts.

A strict conflict-of-interest law covering civilian and military personnel should be passed. The recruitment of top civilian officials of the Department of Defense from defense industry should be prohibited, except in extraordinary cases. There are others available with equal talent and greater objectivity. Nor should military officers be able to count on retiring on an annuity from a defense contractor.

The Defense Department should be prohibited from contracting with universities for war research. Funds now disbursed by the Pentagon to universities for basic research in the physical and social sciences should be transferred to the National Science Foundation or some other civilian agency. War research corrupts the purposes and values of universities and renders the nation's institutions of learning dependent upon the military. If universities need federal subsidies to survive, then the funds should be so labeled and made available from non-military sources.

Essential research functions can be performed in in-house laboratories run by the government on a fraction of the present military-research budget. About 80 percent of the Pentagon's research activities, including research on chemical and germ warfare, war-game scenarios, and other refinements in the science of death, is expendable. As Admiral Hyman G. Rickover told a Senate hearing on the Pentagon's $48 million social-science budget, "No harm would have been done to the Republic if none of it had been done." Neither statecraft nor learning is much advanced by studies on "Witchcraft, Sorcery, Magic, and Other Psychological Phenomena and the Implications on Military Operations in the Congo," an item from the 1968 budget.

The most critical aspect of the military-industrial complex is of course the relationship between the Pentagon and the top defense contractors. That relationship must be radically altered if the society is to move away from further militarization. It is not hard to see what is wrong with the relationship, but improvising solutions is difficult. There are two characteristics of industry's role in the defense process which are highly objectionable. The first is that a relatively few firms derive unfair economic advantages from what amounts to a public franchise. Such firms, which play so vital a role in setting national priorities, are public in every sense but one. They pocket substantial private profits, but are wholly unaccountable to the taxpayers who finance their operations.

The second objectionable feature is that the dependence of the defense industry on the Pentagon gives the military immense added power. It is not true that a few firms control the Pentagon and force it to foment wars to keep them in business. That is the straw-man theory of the military-industrial complex. The Pentagon's ability to control a major share of the nation's industrial production is the real problem, for it gives the Department of Defense a commanding position in the economy which carries with it a dangerous concentration of political power.

A traditional slogan of populist groups in America in earlier days was "Let's take the profits out of war." After World War I even the American Legion endorsed this principle. Recently J. K. Galbraith has proposed nationalizing any firm that does more than 75 percent of its total business with the Defense Department. Taking the steam out of the weapons-pushing process by removing the profit incentive is an attractive notion. The conversion of a small number of absolutely essential defense plants into government arsenals makes sense. Unfortunately, nationalization does not solve the more general problem of concentration of power. H. L.

Nieburg of the University of Wisconsin has put the problem in what seems to me to be exactly the right light:

> The old dichotomy between what is private and what is public is in the process of abandonment; but an explicit rubric which will protect democracy and the public interest has not been formulated. What must come is a system of values and institutions which will replace economic initiative and private property as guarantors of political independence and pluralism. This task of formulation is the greatest challenge of the future. As economic pluralism disappears, only political pluralism, safeguarded by new institutions of representation, can make the exercise of power both responsive and limited. A heightened and more representative infrastructure of interest groups is necessary at all levels of society and may already be forming. The weakness of such interest groups in the past may have contributed to the use of R and D contracting as a form of indirect government intervention in the economy. The problem is not as so many critics of the establishment believe, to control technology; rather, it is to control those interest groups and power coalitions who are—in the name of an automatic impersonal urge toward technological change—making public policy for the nation and holding in their hands much of the power of decision making for the whole society.

If General Dynamics became a government agency, the Pentagon's accountability would not be automatically improved. There would still be no one representing the public interest to ask why the agency should be making a new weapon merely because it is technically feasible to do so. There would be no representative from the local community to point out that the addiction of the local economy to defense spending grows worse with the injection of each additional dollar. The problem of secrecy would remain and would probably be worse.

The hybrid public-private corporation is a major new phenomenon of American life extending far beyond the defense area. Its implications for American democracy are not understood. The Congressional Conference on the Military Budget and National Priorities proposed the creation of a Temporary National Security Committee composed of members of Congress and private citizens to examine the institutional structure of the military-industrial establishment and to make recommendations to Congress. A principal task of such a committee should be to investigate the role of private corporations in setting public policy in the defense area and to propose legislation to subject such corporations to effective citizen control. A possible solution might be to have the public elect a certain percentage of the Board of Directors. The most important step is to reduce radically the flow of public funds to such companies.

How to Give Up the Economy of Death and Keep Prosperity

When it comes to conversion, most Americans are Marxists. They do not believe that the present levels of prosperity or employment can be maintained except by a war economy. They do not see or will not admit the revolutionary implications of their belief. An economic system that works only by turning out products that endanger itself and the planet is literally suicidal. If it is true that American capitalism cannot function without military socialism, then anyone who cares about survival cannot be a capitalist. But is it true?

A central problem concerns unemployment. According to the University of California sociologist Jeffrey Schevitz, one in five jobs in the United States depends directly or indirectly upon the Department of Defense. The Pentagon has 3.4 million members in the armed forces and 1.3 million civilian workers spread across seventy countries. There are 3.8 million industrial workers whose lives are wholly tied to war production. Millions more are indirectly dependent upon the defense budget. Twenty-one percent of skilled blue-collar workers are on military payrolls. Schevitz notes that nearly half of all scientists and engineers in private industry work in the aerospace and defense fields.

These figures give an inadequate picture of the extent to which the nation is addicted to the Economy of Death. Junction City, Kansas, is in the Fort Riley business. It is typical of numerous Pentagon company towns spread across the country. According to *Time* Magazine, when one Army division left Fort Riley in 1965, business fell off 30 percent. Fayetteville, North Carolina, is another string of used-car lots, army-surplus stores, drive-in movies, and more ancient forms of Army entertainment for the diversion of thousands of paratroopers. Without Fort Bragg, Fayetteville would face bankruptcy. Sunnyvale, California, is typical of numerous communities throughout the United States which have become addicted to the war business. From a population in 1940 of 4373 the town has grown to 95,000. Twenty-one thousand work in the Lockheed plant, 3000 for Westinghouse, and 1400 for a division of United Aircraft Corporation. Thousands more work for smaller subcontractors. "If the government says OK, we're not building any more Polaris or Poseidon missiles, and everybody at Lockheed was sitting on the streets, it would hurt real bad," Dan Wood, a Sunnyvale Lockheed employee who averages $10,000 a year helping to build missiles, observed to a reporter from *Newsweek*. Defense contracts and Defense Department payrolls account for more than 20 percent of the entire personal income of such states as Alaska, Connecticut, and Idaho, and the District of Columbia. In California, Kansas, Arizona, and New Mexico between 20 and 30 percent of the industrial employees work for the Pentagon. In Utah the military is the state's number-one employer. When the Army killed 6400 sheep at Skull Valley in 1968 by accidentally spraying them with VX, a deadly nerve gas, the state veterinarian, D. Avaron Osguthorpe, observed, "We've got a defense business bringing in $35 million a year into the state; sheep bring in one thirty-fifth that amount. Which is more important for Utah?"

In its approach to conversion, the federal government has specialized in positive thinking and minimal action. The President's Committee on the Economic Impact of Defense and Disarmament, created in 1965 as a response to rising Congressional concern about conversion, concluded that "even general and complete disarmament would pose no insuperable problems." The Report painted a rosy picture of defense plants as "gigantic job shops" which could easily compete for and fulfill "large-scale non-defense research and development projects." The reality has been less rosy. A few years ago Edmund "Pat" Brown, Governor of California, the number-one state in defense contracts, encouraged aerospace companies to go into research in the fields of transportation, air pollution, and communications. The results were not encouraging. The companies did not have the right skills or adequate incentive. When North American Rockwell is asked to develop a new plane, the Pentagon advances the company enough cash to hire from 70 to 100 professionals. The same company received a contract for a transport study with money for 3 to 5 professionals.

Unlike war work, contracts in the Economy of Life are not riskless, and they do not hold out the promise of big profits for sophisticated hardware. "If the Government provided the risk capital and if the vehicle contained enough technology or patentable elements so that we could close out other producers and if we could market it, then it would be viable," John R. Moore, the president of North American Rockwell's Aerospace and Systems Group, recently laid down as his conditions for a serious commitment to production for peace. Only when making high-speed transport systems is as profitable as making missiles will the weapons-makers voluntarily move out of the death business.

Because of the resistance of aerospace companies to conversion and other factors, official optimism about unemployment has proved to be unfounded. Over 290,000 engineers and scientists work directly in war industries. In 1963–64 because of the completion of a number of military programs more than 30,000 of them lost their jobs. On an average they remained jobless for more than three months. Between April 1963 and December 1964 the Republic Aviation Corporation's Long Island plant laid off 13,600 employees because of a cut-back in the F-105 fighter-bomber. About 16 percent of the men and 36 percent of the women who stayed in the community had yet to find a job more than a year later when the U.S. Arms Control and Disarmament Agency conducted a survey. Of those who did find work, 21 percent of the men and 41 percent of the women were jobless for six months or more. The Boeing Company laid off 5000 workers between December 1963 and March 1964. Despite the fact that these workers were younger and better educated than average, 22 percent of the men and 59 percent of the women were still unemployed in August 1964. According to Schevitz's calculations, new jobs in war industries accounted for 44 percent of the over-all increase in employment in the year 1965. Thus the economy is even less able to absorb defense workers in civilian production than it was five years ago. Each year the problem gets worse. There are two reasons for this. First, salaries payable in defense industries are substantially higher because the firms are able to pass the cost along to the taxpayer. Paying a premium to recruit and keep the best scientific talent in war work has been an explicit national policy. As a result, government-subsidized salaries have risen sharply. Second, military research and development on the "frontier of knowledge" tends to be more challenging professionally than much civilian technology which has not been pushed to the frontier for lack of money and energy. With billions to spend, the Defense Department can set engineers and scientists to exciting tasks that no civilian industry can match. Most engineers and scientists would be able to find work in America's still booming economy, but they are not likely to find work at the same pay or with equivalent challenge. Unless the federal government will subsidize the technology of peace as it has subsidized the technology of war, we face the prospect of a new class of $200,000-a-year hard-core unemployed.

What can the federal government do about unemployment once it is decided to convert the economy from war preparation? The question is central, for unless there is a program instead of platitudes for dealing with this problem, the resistance to conversion will be too strong to overcome.

A national conversion program should be established which operates on the basic principle that the community, not the individual war worker or soldier, must pay the costs of conversion. Like the GI Bill of Rights, such a program would ease the transition for persons whose jobs are destroyed as a result of government policy. Government and a specialized industry have placed millions of workers in a position of dependence on the Economy of Death. They must now bear the responsibility for easing the pain of withdrawal.

Such a program should be administered by a National Conversion Commission with broad powers. A principal job of the Commission would be to assist the retraining and relocation of people released from war research and war production. Under the Manpower and Development Training Act of 1962, limited funds are presently available for "brief refresher or reorientation educational courses in order to become qualified for other employment" to assist people "who have become unemployed because of the specialized nature of their former employment." This program in greatly expanded form should become a major instrument of conversion. In accordance with the national conversion plan, the Commission should award substantial grants to scientific and technical personnel to encourage them to apply their talents to priority problems. The manpower specialist Herbert Striner of the Upjohn Institute estimates that it would take no more than three or four months to train most scientists and engineers now in war work for useful alternative jobs. For example, an engineer who specializes in the miniaturization of electronic components for a missile might be given a federal grant of up to a half-year's salary to study the technology of mass transportation, pollution, or low-cost housing. Another might be given a grant to design an experimental school.

The training programs could be carried out in several different ways. One possibility would be to subsidize on-the-job training programs in civilian plants for former war technologists. The federal government might supplement the salaries of scientists and engineers in non-defense industry to bring them somewhat more in line with the salary scales prevailing in war industry. Firms which have been found to have made excessive profits on defense contracts should be required to set up retraining programs without additional federal subsidy, for they have already received it. Another approach would be to sponsor a program of research and training in technical innovation for the civilian economy at universities. The program would be specifically designed to stimulate innovation in the civilian economy, and technologists from war industries would be given first chance to participate. The universities could run a series of programs ranging from brief refresher courses to full-scale professional schools for new careers. The Commission should also have substantial funds rescued from the defense budget for scholarship grants. Engineers and scientists who make use of these programs should receive tax relief and outright subsidies where necessary. Some scientists and engineers released from defense plants should be encouraged to take federally subsidized sabbatical leaves in which to think about useful alternative ways to spend their lives. For example, some might receive a grant to enable them to teach science in public schools or to set up experimental teaching programs to acquaint adults as well as high-school students with technical and social problems of science. In Washington, the Institute for Policy Studies has sponsored a Neighborhood Science Center that makes it possible for boys and girls in a ghetto area to walk off the street and conduct simple scientific experiments with the help of a skilled scientist. The Commission should make it possible for persons released from the Economy of Death to try similar experiments and to invent others. Such investment in human resources would pay dividends for the whole society.

The same programs should be available for non-skilled workers. In some respects their problem is easier than that of the scientific and technical people. In others it is more difficult. Leonard A. Lecht of the National Planning Association estimates that if $20 billion were cut from the defense budget, half of which went to social programs and half into private pockets through tax reduction, new job openings would exceed jobs lost through defense cutbacks by 325,000. But the new jobs would demand different skills—fewer machine operators and engineers, and more service personnel, craftsmen, and laborers to work in the building trades. The Commission should have a nation-wide program of job placement, retraining programs for unskilled workers, and the power to operate public-works programs with former defense workers.

The Commission should adopt an explicit policy of minimizing relocation to the greatest possible extent. Today it is common for workers laid off in one defense plant to move to a similar job in another city. However, some relocation will obviously be necessary, and the Commission should have funds to purchase houses of defense workers who cannot otherwise sell them at a fair market price. It should also pay travel costs and other relocation expenses. Britain has had a comprehensive program since 1909 to help its workers relocate to find suitable jobs. There should also be an income-maintenance program for defense workers to supplement normal unemployment benefits. Defense plants should be required to contribute to such a program by establishing insurance funds out of excess profits.

Communities that are dependent upon war industries or military installations should be eligible for special assistance. In a real sense they are "disaster areas" and should be eligible for the sort of extraordinary relief that is given to communities stricken by flood or tornado. When the Studebaker plant in South Bend, Indiana, closed in 1964, resulting in the layoff of 8700 workers, a coordinated federal program of assistance was undertaken. However, the principal instrument used to rescue the community was the defense budget. The Department of Defense arranged for the sale of the Studebaker plant to other defense firms. The Commission should stimulate non-defense production on a crash basis by making low-interest loans and emergency grants to affected communities. It could turn over federally owned military installations to the local community at nominal cost upon receipt of a community plan for the utilization of the property. In some cases an emergency negative income tax or other income-maintenance program would be necessary.

The costs of these and other programs that might be developed by the Commission would be considerable, but they would be nowhere near the savings that could be realized by a substantial cut of the defense budget. Further, every dollar spent to stimulate useful production of goods and services needed to prevent social decay and to remove the injustices that lead to violence contributes to the real wealth of the nation as weapons stockpiles do not.

Planning an Economy of Life

The major task of the National Conversion Commission would be to prepare a national conversion plan. For a generation, military spending has fulfilled three social purposes in America quite apart from defense. First, it has served to distract us from domestic problems. In the 1950's, according to the standard rhetoric, the nation could afford a military force "second to none" and do everything that needed to be done at home. (The unstated premise was, of course, that there wasn't much to be done.) In the 1960's, when it was no longer possible to deny the problems of the cities, the response was to blame unavoidably high defense costs for the national failure to meet the crisis. Second, military spending has satisfied the Keynesian economists' demand for a high level of government spending to stimulate growth. Third, by having $45 billion a year to spend in the economy, the Pentagon has become the principal planner in American society.

If the federal government is to take the lead in reordering national priorities, it will have to exercise openly and rationally two functions that it now exercises covertly and irrationally. The first is subsidization, and the second is planning. A widely believed economic fairytale has it that there is an invisible wall separating

the public government and private enterprise, that men
get rich in spite of the government and not because of
it. The reality is otherwise. The United States is a highly
subsidized society. Not only defense contractors but oil
interests, construction interests, shipping interests, and
many others receive billions of dollars' worth of subsidies funded by the taxpayer in the form of depletion
allowances, administered prices, and cash benefits.
There is nothing wrong in principle with subsidies. The
American system probably could not function without
them. The important political questions are: Who gets
subsidized? What national purpose is served? Are the
taxpayers' interests protected?

There are strong traditions in America against using
the federal budget to subsidize the public sector of the
civilian economy. To take tax dollars to buy schools,
hospitals, or food within the United States in adequate
amounts is enormously difficult under our system. To
buy weapons to fight an external enemy, on the other
hand, is easy to understand and, until recently, almost
impossible to oppose. It was not hard to find $687
million for germ warfare or $70.8 million for defoliat-
ing Vietnam in 1968, but it took a major struggle to
obtain $100,000 to find out who goes to bed hungry in
America. As one top weapons manufacturer put it recently, "We have the politically saleable animal—much
easier than Watts."

To invest in the reconstruction of American society
means redistribution of wealth and power, while military socialism for the most part means subsidizing the
rich. Spending a billion dollars in the defense economy
can be done with a few telephone calls and lunches
involving a small number of people with an almost
classic harmony of interests. Spending a billion dollars

to stop pollution in New York or to fight poverty, on the other hand, involves a collision with entrenched political forces interested in the status quo. It is even easier to spend a dollar on the moon or on the bottom of the sea than on the poor, the hungry, the sick, or the old in America's cities or on her farms. Under the Puritan ethic, the government, like God, helps those who have helped themselves. For a politician, investing in the Economy of Death is bipartisan statesmanship. Investing in the Economy of Life involves a near-certainty of making at least one political enemy. To survive without civil war, however, America must now subsidize the "losers" instead of the "winners" in the national success race. Only a just society can dispense with violence.

To subsidize the Economy of Life would take some fundamental changes in American political attitudes. How these changes might come about will be discussed below. It will also take planning. This is another concept that traditionally has been greeted with great suspicion in American life. Much of what happens in our society is the result of private decisions and public non-decisions. It is clear, however, that if the society is to shift a substantial portion of its investment in military socialism to other public purposes, the reallocation of those resources cannot happen without some fundamental decisions of a public character. The problem is not finding things on which to spend the money. More than six years ago, even before the nation had discovered the true dimensions of poverty in America or the environmental crisis, the Columbia University economist Emile Benoit calculated that the nation needed to spend between $65 billion and $77.4 billion a year on itself to assure a decent society. The costs for the 1970's will be substantially higher than this. The prob-

lem is the lack of effective institutions for investing well
in American society. High military spending has di-
verted attention from this problem for a generation.

The National Conversion Commission should have
the responsibility to prepare a national conversion plan.
As a first step it would, in cooperation with other
domestic agencies and departments, put together an
inventory of national needs. How many units of housing? How many medical schools? How much of an
investment in air-traffic safety? What would it take to
clean the nation's rivers? What would it take to feed the
hungry? What would it take to make black people in
the ghetto economically independent? What would it
cost to purify the air? The Commission would attempt
to determine what was needed for these tasks in manpower, facilities, and money. Each year the Commission would publish a State of the Union message on
national needs and how far the nation was from meeting them.

Next, the Commission would request each community in the United States to prepare its own "wish list"
or inventory of local needs. How many new beds for the
hospital? How many schoolrooms? How many new jobs
are needed for the rising population and what kind?
What would the people of the community like to make?
What would they like to do? Every community in the
United States with a substantial dependence on military
spending would be required to prepare a local reconversion plan as a condition of receiving further federal
grants of any kind. Each defense contractor would be
required to list the number and types of jobs presently
dependent upon defense work and to submit a conversion schedule. The schedule would show what alternative products the contractor was prepared to make,

what percentage of his present labor force he could employ, and what federal aid, if any, he would need to assist in training or relocating his employees. Defense plants would be given economic incentives to shift to the production of items listed on the local "wish list."

The National Conversion Commission would face one major dilemma. While a national plan and inventory of needs is essential, centralized administration of the conversion program would be a disaster. To replace the Pentagon bureaucracy with a pollution-control bureaucracy of similar size and character would increase the chances of survival but not the prospects for freedom. A major source of the public unhappiness is frustration with elephantine bureaucracies that literally crush the programs they supposedly administer. It has become a cliché to observe that people want greater participation in making the basic decisions that affect their lives, but the objective has seldom been realized. The conversion process should be regarded as an opportunity to experiment with new political forms for making decisions about allocating resources. Town meetings and city councils should debate the national conversion plan as well as the inventory of local needs. The National Conversion Commission should have the power to direct grants to units modeled on the "ward republics" proposed by Thomas Jefferson or the ECCO project developed by Milton Kotler in Columbus, Ohio, in which the neighborhood elects a government to look after local interest. Neighborhood health clinics, development banks for ghetto industry, local cooperatives, and other experimental structures for the society should be tried. Programs for economic development, agricultural reconstruction, and combating water pollution may need to be carried out on a multi-state basis, and the Commission's powers should be broad enough to encourage such regional structures.

What Can the Citizen Do?

The task of shifting American priorities is primarily a political rather than an economic problem. As we have seen, the federal government must play an active role in redirecting investment into essential areas of the public economy and in helping individuals and communities now dependent upon war industry to play a productive role in a life-oriented society. Most economists are confident that the economics of conversion could be successfully managed and that the nation as a whole would emerge from the process richer and better equipped to solve its problems. Between 1944 and 1947 the defense budget was shrunk by almost 90 percent with no loss of prosperity. At the height of World War II, defense spending accounted for 41 percent of the Gross National Product. It is less than 10 percent today.

The problem is one of will and incentive. The Department of Defense and its captive industry have achieved a privileged position in American life, enjoying substantial exemptions from the rules that govern the rest of us. They need not explain what they do. There is no great necessity to tell Congress or the public the truth. "Promise anything and charge twice as much" has been the operating creed in defense procurement. Generals and Pentagon administrators do not look forward to the day when others will decide how to spend the money that they have so long controlled. It is too much to expect giant bureaucracies to greet their own liquidation with enthusiasm. One should not expect to see the executives of General Dynamics at Quaker peace vigils.

Nevertheless, the military-industrial complex has a hold on American society which must be broken. Nothing less than the survival of the nation is at stake. How is it to be done?

There are four principal groups in American society which are becoming aroused by the dangers of American militarism, each for slightly different reasons. They constitute the cadre of any peaceful revolution to change American values and American priorities.

The most active group is made up of the students. Faced with the existential crisis of deciding between killing in Vietnam, spending three to five critical years in a federal penitentiary, or going into exile, the generation now in the university is in moral and intellectual crisis. Growing numbers of the most articulate and, by the universities' own standards, brightest students can no longer stand the "system." They do not understand why medical schools, such as the University of California, should do research on germ warfare. They do not see why they should be asked to kill Vietnamese or Dominicans in the name of national security. They are outraged at leaders who appear incapable of leading the nation any place but into war. They are the first generation who have heard from childhood what nuclear war means. Having missed much of the apocalyptic rhetoric of the early Cold War period, they are not worried about the Vietnamese or Cubans taking over the country. They are more concerned about the only country in the world that has consistently pursued victory in the nuclear arms race, that has sold or given away more than $50 billion in arms to other countries in a little over twenty years, that maintains a string of bases on every continent, and that has rained more bombs on the small country of Vietnam in two years than the Allies dropped on all of Europe in the five years of World War II. These facts suggest to them that it is the United States more than Russia which needs to be contained.

When the student editors and student presidents of the most distinguished universities in the country resist the draft and express open contempt for the nation's leaders, it is clear that far more than a radical fringe is involved. Indeed, the issue that divides the revolutionary students from the moderates is not whether there is a military-industrial complex or whether the U.S. is a Warrior Nation, but whether there is hope for the present American system. Can America survive except as an empire? Is reconstruction possible without bloody revolution? Moderates believe that the American system can shift priorities and adapt to the obvious needs of the time, and the revolutionaries believe that the state must be destroyed before the people can be saved.

Because students have been invited to kill and to die under the present national-security system, they have had considerable incentive to take a good look at it. Many have performed an important national service by exposing the secret involvement of their universities in war research. Groups of students have begun to investigate specific operations of the Pentagon, much as Ralph Nader has done in other areas. Others have been looking at local war industries and their role in the local community. Such research is a vital prerequisite for a public education campaign on the militarization of America and the distortion of priorities. A seminar on the military-industrial complex is being given at Yale. Such courses should be demanded by students at every

university. Political-science curricula that do not discuss this major phenomenon of American life are per-
petuating illusion instead of teaching truth.

Students have the intellectual and moral energy to
confront America with what we have become. Systematic exposure of what men and institutions are doing to
ourselves and to other people in the name of security
can touch the conscience of the community. There is
about the military-industrial complex the same American dilemma that Gunnar Myrdal found in the race
problem, a profound tension between animal fear carefully cultivated by demagogic politicians and an innate
sense of decency and fairness. As in the race problem,
conscience alone is not a sufficient force for essential
social change. But it is a necessary beginning to effective political action. The psychological defenses that
support military socialism must be stripped away. Students, who are trying to find out who they are, can help
older Americans rediscover themselves and their country. Without the encumbrances of success, they are in a
position to resist serving the Warrior State and to refuse
the positions reserved for them in university war research and in weapons production. By their example
and exhortation, they can help others to think about
whether they should spend their lives in the preparation
for war.

Another group that is becoming mobilized against
the distortion of national priorities is made up of some
of the nation's leading scientists and technologists who
understand better than laymen that there is no technical
solution to the arms race. Many scientists and engineers
in war industry and in university war research are torn
by the contradiction between the intellectual excitement
they feel in being on the frontier of technology and their

guilt at the use to which their knowledge is put. At Massachusetts Institute of Technology, younger faculty have taken the lead in the campaign to cut the dependence of that institution on war research. When university scientists are willing to stop cooperating with weapons production and war research, the juggernaut will be slowed.

M.I.T. scientists have also begun to work with students interested in studying the local embodiment of the military-industrial complex in the Cambridge area to help that community understand its dependence on the military. Much of the recent outrage about military waste and excessive power of the military grew out of specific revelations about the activities of the military-industrial complex. When, for example, Congressman Richard McCarthy announced early this year that by 1954 the Army had produced enough nerve gas to kill everyone on earth 1000 times over and was planning to ship 11,000 tons of it across the country on railroad cars, the public learned far more about militarization of American life from this glimpse of the Pentagon underworld than they do from the most eloquent abstract speeches. Continued exposure is essential not only to curb the excessive power of the Pentagon which thrives on secrecy, but to overcome public apathy, its most powerful political ally.

An issue which has become important on many campuses is whether the war-research laboratories should be owned and operated by the university or should be independent "off-campus" operations. Severing the university's ties with such research laboratories may help the university to recover its innocence, but it does not materially affect the pace of war research if the same scientists continue to work and receive their checks from an independent military contractor or from an "in-house" laboratory of the Pentagon. The scientific community has considerable political power. It should mount a campaign in its own interest as well as the national interest for a national research policy that will favor innovation for the solution of critical public problems. Such a policy would necessarily mean drastic cutbacks in military research, for there is neither money nor talent to do both. In support of such a campaign, scientists could undertake a critical educational campaign to explain to the public that there is no technical answer to security. There are no absolute weapons. Every addition to the arsenal can be offset by the adversary. Public understanding of these issues is essential if the people are to protect themselves from the scare tactics employed by the Pentagon.

At the same time, the scientific community should continue the efforts already made to organize scientists and engineers in war industry to press for conversion. If within each plant there were a group of engineers willing to confront the management with its failure to take steps toward conversion, significant counter-pressures to the military-industrial complex would grow. Scientists who are concerned about the hold that the Economy of Death has on the nation's engineers might organize a placement service to help those who want to leave war industry to find alternative employment. A union of scientists called the Technology and Society Committee has been formed in the San Francisco Bay area, one of the heaviest concentrations of war-related industries in the country. Their mission is to organize war technologists in the area to encourage them to change the character of the institutions for which they work or to assist them if they leave. Professor Jeffrey

Schevitz of the University of California has documented the growing dissatisfaction of scientists and engineers with war preparation as a career. The Rand Corporation, which ten years ago enjoyed great prestige with top graduates of leading universities, now has trouble, according to its president, Henry Rowan, recruiting people for war-related work. Schevitz interviewed a disillusioned engineer whose attitude is becoming increasingly common:

> At this time I found myself working on a project analyzing some electronic equipment to be used in Vietnam. And I found it was very enjoyable for me analytically, but I began to be more and more troubled about the nature of my work, especially when I found myself sitting next to my desk with a map of North Vietnam, trying to figure out the optimal route for aircraft to take to come in and bomb cities without getting shot down. That hit a little too close to home, and it forced me to think about a lot of things that I hadn't thought about before and finally I got to the point where I just could not, in clear conscience, continue to work on the project. I had to take a stand either for or against the war at this point. I made my decision and shortly thereafter quit the project.

A third group in the nation with considerable potential for helping to create the political climate for new national priorities is made up of businessmen who are not Pentagon clients. More and more of these businessmen are questioning the myth that high defense expenditures ensure prosperity. They are worried about inflation that eats savings and pushes up interest rates. The managers of non-war-related industry find that they cannot get the capital for expansion or the scientific talent to innovate in such fields as transportation and pollution control. Professor Kenneth Boulding, testifying before the Proxmire Committee, put it this way:

> There is a great deal of evidence to suggest that civilian industry is starved of able research scientists and engineers because of the "internal brain drain" into the war industry. By far the most important resource of any society from the point of its future is the innovative capacity of its ablest minds, and if these are absorbed into the war industry they will clearly not be available elsewhere. Furthermore, the spillovers from the war industry into the civilian sector seem to have been declining, and are quite inadequate to compensate for the drain of problem-solving capacity. Because of the obsessive expansion of the war industry many vital sectors of the civilian economy are failing to solve their technical problems.

Marriner S. Eccles, former chairman of the Federal Reserve Board and prominent investment banker, summarizes some of the major concerns of businessmen about the militarization of the economy:

> I am convinced that business generally, in the long run, is being injured by our over-kill spending of the military. It is directly responsible for the dangerous inflation that has been developing, particularly in the past two years, as well as the huge and increasing deficiency in our international balance of payments which is tending to destroy the dollar as the principal world currency; also the record high interest rates and increasing taxes. It is also

responsible for our financial inability to adequately meet the problems of our cities (poverty, crime, riots, pollution) and our rapidly expanding educational requirements.

The well-known investment analyst Eliot Janeway shows the impact of defense spending and the Vietnam war on the stock market:

Thanks to the Vietnam crisis, the first market crash since the coming of the welfare state to America became a clear and present danger. . . . When the anticipated benefits of a small war and a big boom turned into the actual burdens of a big war and a costly squeeze, the economy was whipsawed between the inflation of costs and the deflation of earnings. The stock market was exposed to the shock which recognition of a war crisis always brings. Not until the Vietnam crisis had taken its toll of the money market, of the earning power within the economy and of the stock market, did either opinion or the government awaken to a realization of how far out of control the financing of the war was or how dislocative its economic consequences had become. By that time, America had reverted to the old European practice of making war at the expense of the economy.

A longer-range economic cost is the degeneration of non-military technology. The United States is first in the world in military optics. We have developed reconnaissance cameras that can photograph a golf ball at 70,000 feet. But the nation is behind Japan, Germany, and other countries in producing microscopes and cameras. We have the world's best lasers, which fascinate the military because of their possible use as death rays, but the technology of traffic control, low-cost housing, and mass delivery of health services is seriously lagging. Japan, which has virtually no military expenses, is far ahead in mass transport. Scandinavian countries are ahead in housing. The *Wall Street Journal* reported almost seven years ago:

> Frantic bidding by space and military contractors for scientists and engineers is creating a big shortage for industry. This scarcity, along with the skyrocketing salaries it is provoking, is bringing almost to a halt the hitherto rapid growth of company-supported research. . . . Samuel Lebner, vice president of Du Pont Company, puts it this way: "Government research programs serve as a brake on research in the private sector."

Industrial research is the wellspring of domestic productivity. Men in industry are worrying that it is begining to dry up.

Businessmen are particularly worried about high taxes that are directly related to the size of the defense budget. In addition, they are feeling the effects of increasing local taxation that is made necessary because of cutbacks in federal grants to states and local communities. In specific industries such as housing, high interest rates are having a destructive effect. Moreover, non-military industry must keep raising wages to compete with industry without receiving the substantial subsidies available to Department of Defense contractors in the form of government-supplied capital, plant, and facilities. The advantages to a few favored defense corporations is causing increased dissatisfaction in the business community.

Some businessmen have formed groups to oppose the Vietnam war and excessive military spending, such as Business Executives Move for Vietnam Peace, the Businessmen's Educational Fund, and the Fund for New Priorities. Others are active in political groups organized to fight the ABM or on other military spending issues. The business groups can play an important role in alerting the rest of the country to the economic risks to national security in devoting a disproportionate share of national resources to inherently wasteful military expenditures. They have the credibility of success. Normally, they do not ask whether spending so much on death is morally right, but whether it is fiscally sound. This places them several notches higher in public esteem than preachers and intellectuals.

Such business groups can play an important role in persuading the country to take the conversion problem seriously. They can interest local financial and commercial leaders in analyzing the extent of their community's dependence on war industry and in thinking seriously about how to run the local economy on something other than war preparation. Businessmen might take the initiative in organizing local Committees on Community Development which would assess local needs, prepare local conversion plans for each local industry, and lobby for federal aid. In short, they might anticipate on a voluntary basis what I have proposed as a required activity under a conversion law. By acting as if a national conversion policy were already adopted, they can help create the political climate to bring it to reality. The stronger the participation of local businessmen, the greater the pressure that can be exerted on defense contractors and subcontractors themselves to take steps toward conversion. If the local community becomes aroused about the morality, necessity, and economic wisdom of the local napalm plant, as the citizens of Redwood City did, the managers of war industry may begin to take the threat of peace seriously and plan for it instead of against it.

Businessmen can also play an important role in the educational campaign necessary to convert the country to an Economy of Life. They should appear before civic groups, ask to testify before town and city councils, and press the local universities to consider the economic and social issues raised by conversion. Specifically, they should lend their prestige and respectability to the puncturing of economic myths about America, and should demand a rational program for subsidizing the Economy of Life in which the decisions are publicly debated and the recipients are publicly accountable. They should oppose higher taxes unless these are specifically linked to reduced defense expenditures. They should support, morally and financially, the students who raise fundamental issues of militarism in a serious way and expose the military underworld.

The fourth group is potentially the most powerful. Members of Congress have recently come to see that they must vigorously reassert their constitutional power over military spending and foreign policy or lose it irrevocably. The erosion of Congressional power in these vital areas is already far advanced. Most Congressmen have been willing to take the Pentagon's word on almost everything because they have seen no alternative. Senator Gale McGee admits publicly that he learns what he needs to know about new weapons from the "Huntley-Brinkley Show."

One sign of new Congressional attitudes is the report of the Senate Foreign Relations Committee issued in 1967:

> There is no human being or group of human beings alive wise and competent enough to be entrusted with such vast power. Plenary powers in the hands of any man or group threaten all other men with tyranny or disaster. Recognizing the impossibility of assuring the wise exercise of power by any one man or institution, the American Constitution divided that power among many men and several institutions and, in doing so, limited the ability of any one to impose tyranny or disaster on the country. The concentration in the hands of the President of virtually unlimited authority over matters of war and peace has all but removed the limits to executive power in the most important single area of our national life. Until they are restored the American people will be threatened with tyranny or disaster.

The Senate has passed a resolution calling for Presidential restraint and prior consultation in the making of foreign commitments. The upper house has begun to reassert its old authority because its leading members are tired of being misled by the executive. Senator Stuart Symington, for example, once a leading advocate of giving the military what they ask, has become an outspoken and effective critic. The products of the salesmen of fear in the Pentagon, the bomber gap, the missile gap, the civil-defense gap, and the ABM gap, have not stood the test of time. Symington, Proxmire, and other Senators have played an important role in exposing waste, misconduct, and dangerous practices.

Such hearings should become a regular activity, for not only do they have some restraining effect on the Pentagon but, more important, they have an educational impact on the country and can help create the climate for breaking up the military-industrial complex and shrinking the defense effort to proper size.

Some Congressmen, in connection with the Vietnam war and the ABM fight, have conducted local hearings in their districts to which constituents come to ask questions, to make their views known, to press the Congressman to commit himself on issues, and to be educated on foreign and defense policy. These hearings are a good vehicle for increasing democratic participation and control on these issues. Each year the defense budget should be discussed at a series of such hearings in each Congressional district. Congressmen themselves should take the initiative to set them up. Where they don't, local citizen groups should organize them and invite the Congressman to appear. If a Congressman refuses, the group should organize immediately to defeat him at the next election or even to have him recalled if the state constitution has such a provision, for it is unthinkable for an elected representative to refuse to discuss life-and-death issues of war and peace with his constituents.

Congressmen could perform a great service by educating the public to the reality that the decisions on defense and foreign policy are matters of political rather than technical judgment. They are no easier but probably no harder than many other public policy decisions that have to be made in a complex modern society. Only the stakes are higher. Technical and secret information plays a minor role in the fundamental decisions on how large the defense budget should be. In the

Kennedy Administration, for example, the decision to buy 1000 Minuteman missiles was based not on McNamara's computers but on a political compromise between the Joint Chiefs of Staff, who wanted 2000, and the White House staff, who thought a few hundred were quite sufficient. There are very few magic numbers or critical secret facts that are not printed in *The New York Times* a few weeks after they appear in top-secret memoranda. The most closely guarded secrets concern the details of intelligence-gathering operations. They play only a marginal role in the basic strategic choices of the top National Security Managers themselves and are not essential for an informed analysis of those decisions by an intelligent citizen.

Congressmen have an obligation to become sufficiently informed themselves to be able to discuss intelligently with their constituents what they can know about military spending, what they can't know, and what difference it makes. Only when the issue of secrecy and technical information is confronted directly will the citizen be able to come to a better judgment than "The President knows best." The "If you only knew what I know" mystique undermines the very principle of democratic government. Unless it is publicly and consistently challenged, the American system is threatened.

Congressmen should demand far greater access to information than they now have, and should regard it as their responsibility to pass information on to their constituents. Secrecy should be constantly challenged in Congress, for it is used more often to protect reputations than vital interests. There should be a standing Congressional committee to review the classification system and to monitor secret activities of the government such as the CIA. Unlike the present CIA review committee, there should be a rotating membership. The risk to the nation of compromising classified information is far less than the risk of an invisible government protected from the people by a semi-permanent group of friendly overseers.

The crucial power of members of Congress is, of course, the vote. It cannot be emphasized enough that this is the most direct and effective instrument for shrinking the Department of Defense. When Congressmen vote no on military appropriations, the Defense Department will come up with a national-security strategy that costs significantly less. But not before.

Other groups in American society are also beginning to play a part in conversion. Many religious denominations have refused to bless the Vietnam war, and individual clergymen have witnessed against the legitimacy of American military policy. When all the churches and synagogues refuse to lend their support to a foreign policy and an economy based on the accumulation of killing power, the prospects for conversion will improve. Individual churches should set up study groups to go over the defense budget. Each year they should send for copies of the Secretary of Defense's Posture Statement. A request for 50 million copies would itself have considerable political significance.

Perhaps the most important contribution clergy can make is to explore with their congregations what it is in their spiritual biographies that makes them look to violence for security. The most profound obstacles to conversion are psychological and spiritual. What is it about the American people that makes us fear uncertainty? Why are we so ready to shower honors on the top dog, whether in the local community or in the global community? What can we do about the will to dominate which stimulates destructive competition? It is often said that "human nature" is the cause of war, and books like Konrad Lorenz's *On Aggression,* which are interpreted to support that view, are very popular. But it is individuals with names who make the decisions to prepare for war and to go to war. The churches should reassert their prophetic role of naming names. If a man holding high office is a war criminal, judged by the standards of the World War II Nuremberg trials, in that he advocates launching a nuclear war, condones the torture of prisoners, or sanctions the indiscriminate bombing of civilians, then someone should say so. Given their traditions and professions, the churches cannot in conscience evade this unpleasant responsibility. Nevertheless, the more difficult task is not identifying villains but discovering what it is in ourselves that makes us willing to be accomplices.

Labor unions have been among the strongest forces in the country in support of the Economy of Death. The major unions maintain lobbyists in Washington to make sure that defense contracts continue to keep their members employed. Nevertheless, there has recently occurred in some unions a rethinking of labor's interests in the economy. The United Auto Workers have opposed the ABM and have called for a cutback in defense spending. Nowhere is an educational campaign on issues of security more important than within the labor movement. There should be established in each local a forum for raising and discussing issues of defense spending and their relation to the personal security concerns of workers. Congressmen with labor constituencies should make a special point of talking frankly with union rank and file to build understanding and support for breaking up the military-industrial complex and reducing military spending. Congressmen and other speakers should insist upon the right to speak to defense-plant workers on the premises. The national interest in permitting free discussion on these issues is at least as strong as the interest in permitting union organizing on company premises.

Finally, what about the citizen who belongs to none of these constituencies? He needs to educate himself about the issues of military spending and the role of the military-industrial complex. He can do some of this on his own through reading and listening. He should hound his Congressman if the Congressman is not alert to the dangers of militarism, and give him active support if he is. He should work at the precinct level in persuading local political organizations to take forthright positions against the Economy of Death, on both national and local issues. Individual citizens can help persuade city and town councils to continue the practice begun in the Vietnam war of taking positions on foreign-policy and defense questions. They can also organize or assist door-to-door educational campaigns to discuss issues of security and military spending. Only when the public is sufficiently informed so that it can no longer be stampeded into supporting whatever the Pentagon asks will our leaders abandon this easy tactic.

One important thing a citizen should do is demand that his elected leaders begin to call things by their right names. "Peace is our profession" may be a good slogan for the Strategic Air Command to put on bombers and hydrogen bombs, but the reality is that the Pentagon's only strategy of peace is to offer the country security through killing. Whether that is or isn't the best way is arguable, but the military should be required to state what they are doing. The citizen should demand Truth-in-Packaging from the Pentagon. Whatever else it is, the ABM is not a "building block of peace." It is technically correct to call a nerve gas that literally strangles its victims in seconds with their own muscles a "toxic agent," but it hardly does it justice. "Damage limitation capability" should read "force capable of launching a reasonably successful surprise nuclear attack." "Pacification program" should read "resettlement, internment, and bombardment program." It may not be a coincidence that since the War Department became the Defense Department it has been much more warlike. Citizens who relentlessly demand the truth have taken a giant step toward the recovery of their freedom.

Unfortunately, however, it is too late in the day for knowledge and truth alone to rescue the nation from twenty-five years of militarism. Truth can liberate only if people are prepared to act on it. If they are not, the most shocking revelations have no effect other than to feed the general apathy and to reinforce immobilism. The institutions that make up the military-industrial complex wield their great power because they are able to pose as a legitimate, even an essential, force in American life. But the power and legitimacy of the military underworld rest on widespread public acceptance, and their grip on American society will not be broken until a substantial segment of the public comes to see the military-industrial complex as a serious and immediate threat to national survival and is prepared to confront it as such.

References

INTRODUCTION

See the remarks of Congressman Robert Leggett in the special issue of *Progressive* magazine (June 1969) for explanation of the real extent to which military spending takes up the tax dollar.

For a recent authoritative discussion of the environmental problems alluded to in the text, see Hearings, Senate Committee on Interior and Insular Affairs, April 16, 1969.

CHAPTER I

All figures on U.S. and Soviet force levels come from the annual Posture Statement of the Secretary of Defense, which can be obtained from the Department of Defense, Washington, D.C. 20301. It is a document of 150–200 pages which gives the rationale for the weapons systems and troop strength demanded. These requests are frequently characterized in the document as "austere" and "modest."

Former Budget Director Charles Schultze's discussion of probable trends in military spending can be found in Kermit Gordon, ed., *Agenda for a Nation* (Doubleday, 1968). In the same volume, see also an article by Carl Kaysen, former Deputy Special Assistant to the President for National Security Affairs, on how to cut the defense budget by $30 billion.

Accounts of missile and bomber gaps and other security scares can be found in Samuel P. Huntington, *The Common Defense* (Columbia, 1961) and Paul Hammond, *Organizing for Defense* (Princeton, 1961).

Congressional hearings are another invaluable source of information on military planning. Each year the House and Senate Armed Services Committees hold hearings on the proposed budget. The hearings are closed, but edited transcripts are published several weeks later. The deletions almost always are limited to specific details considered military secrets. Until now neither the committees nor the witnesses have found the inconsistencies and untruths that litter these proceedings sufficiently embarrassing to suppress them. The House and Senate Appropriations Committees also hold hearings on the defense budget. In addition, special committees such as the Senate Preparedness Committee hold hearings from time to time. Transcripts are available from the Government Printing Office, Washington, D.C. *I. F. Stone's Weekly* has performed a valuable service for many years by documenting some of the more egregious instances of carelessness with the truth. Stone should have competitors in this activity. Some news service or newsletter could perform a patriotic service by publishing extended edited and indexed extracts from these hearings for wide public distribution.

For a hair-raising discussion of biological warfare, see Seymour M. Hersh, *Chemical and Biological Warfare* (Bobbs-Merrill, 1968).

For accounts of the Kennedy Administration and the 1961 Berlin crisis, see Arthur Schlesinger, Jr., *A Thousand Days* (Houghton Mifflin, 1965) and Theodore C. Sorenson, *Kennedy* (Harper and Row, 1965). See also William W. Kaufmann, *The McNamara Strategy* (Harper and Row, 1964).

The special statement of the Director of Defense Research and Engineering on research and development can be obtained by writing his office at the Pentagon.

The Nixon-Laird conversation in the text is reconstructed from Laird's testimony before the House Armed Services Committee and the Senate Foreign Relations Committee in support of the ABM and President Nixon's campaign speech at Fort Worth, Texas.

For a discussion of the Soviet Union and the arms race, see Lincoln P. Bloomfield *et al., Khrushchev and the Arms Race* (M.I.T., 1966).

The discussion of the General Purpose Forces is based on Secretary Clark Clifford's final Posture Statement. For an important analysis of the usefulness of U.S. foreign commitments, see Hearings, Senate Foreign Relations Committee, *U.S. Commitments to Foreign Powers* (1967).

For a discussion of U.S. counter-insurgency policy and its political failure, see my *Intervention and Revolution: America's Confrontation with Insurgent Movements Around the World* (World, 1968).

For a discussion of the moral dilemmas posed by revolution in corrupt and backward societies, see Robert Heilbroner, "Counterrevolutionary America," in *Commentary,* April 1967.

For an exhaustive and chilling account of the arms trade, see George Thayer, *The War Business* (Simon and Schuster, 1969).

H. L. Nieberg's *In the Name of Science* (Quadrangle, 1966) is an excellent account of some of the political problems of controlling technology.

Chapter II

Recent attacks on military spending have been best reported in the Washington *Post.*

Accounts of Pentagon profiteers can be found in Clark

Mollenhoff, *The Pentagon: Politics, Profits and Plunder* (Putnam, 1967).

The best sociological study of the professional military is Morris Janowitz, *The Professional Soldier* (Free Press, 1960). An overly cheerful but extremely informative account of military-civilian relations is Samuel P. Huntington's *The Soldier and the State* (Harvard, 1957). To get a real taste of "the military mind," one must read the generals' own literary works: General Thomas S. Power, USAF (Ret.), *Design for Survival* (Coward-McCann, 1965); General Nathan F. Twining, USAF (Ret.), *Neither Liberty Nor Safety* (Holt, Rinehart and Winston, 1966); General Curtis E. LeMay, USAF (Ret.), *America Is in Danger* (Funk and Wagnalls, 1968). See also Walter Millis, *Arms and the State* (Twentieth Century Press, 1958).

There have been a number of good critical studies of the military establishment: John M. Swomley, *The Military Establishment* (Beacon, 1964); Fred J. Cook, *The Warfare State* (Macmillan, 1962); Ralph E. Lapp, *The Weapons Culture* (Norton, 1968); Tristram Coffin, *The Armed Society* (Pelican; originally published as *The Passion of the Hawks* [Macmillan, 1964]). See also the important review article "Is There a Military-Industrial Complex Which Prevents Peace" by Marc Pilisuk and Tom Hayden, reprinted in Robert Perrucci and Marc Pilisuk, eds., *The Triple Revolution* (Little, Brown, 1968). Also Merton J. Peck and Frederic H. Scherer, *The Weapons Acquisition Process* (Harvard Business School, 1962).

Accounts of wartime mobilization and its impact on American society can be found in the U.S. Bureau of the Budget history *The U.S. at War* (U.S. Committee on Records of War Administration, 1946); Eliot Janeway, *The Struggle for Survival* (Yale, 1951); and Bruce Catton, *The War Lords of Washington* (Harcourt, Brace, 1948). On the same point and many others, *The Forrestal Diaries,* ed. by Walter Millis (Viking, 1951), are invaluable. The best accounts of the public-relations activities of the Pentagon are to be found in John Swomley, *Press Agents of the Pentagon* (League Against Conscription, 1953); an article by Walter Pincus on the "Starbird Memorandum," in the Washington *Post;* and an excellent CBS television documentary on the making of the film *Tora! Tora! Tora!*

In 1962 extensive hearings were held on the subject of public political statements by military officers: Hearings, Special Preparedness Subcommittee of Senate Armed Services Committee, *Military Cold War Education and Speech Review Policy* (1969). Some of the material on public relations in this book was borrowed from an original research paper by William Stivers, a student at the Institute for Policy Studies, a debt I am pleased to acknowledge.

Some early instances where civilians outdid the military in pushing preparedness and military intervention can be found in Samuel P. Huntington, *The Common Defense* (Columbia, 1961).

For an historical account of the term "national interest" in the United States, see Charles Beard, *The Idea of National Interest* (Macmillan, 1934).

Bernard Nossiter, in an important and pioneering series of articles in the Washington *Post* during December 1968, has exposed better than anyone else the thinking of the leading military contractors. These articles were based on extensive interviews with top company executives.

In addition to the Report of the Proxmire Subcommittee of the Joint Economic Committee mentioned several times in the text, the reader should consult the hearings that preceded the report (particularly the testimony of Murray Weidenbaum, Admiral Rickover, and A. F. Fitzgerald): *Economics of Military Procurement*, Subcommittee on Economy in Government of the Joint Economic Committee (1969).

In addition to the Phelan article quoted in the text, the reader should consult the forthcoming study *Pentagon Capitalism* by Seymour Melman (McGraw-Hill, 1970).

CHAPTER III

J. K. Galbraith's book is entitled *How to Control the Military* (Doubleday, 1969).

The studies of the Boeing and Republic Aviation cutbacks can be obtained by writing to the Assistant Director for Economic Affairs, U.S. Arms Control and Disarmament Agency, Washington, D.C.

See Kenneth Boulding and Emile Benoit, eds., *Disarmament and the Economy* (Harper and Row, 1963) for a discussion of the over-all impact of substantial cuts in military spending. See also Seymour Melman, *Our Depleted Society* (Holt, Rinehart and Winston, 1965).

The North American Committee on Latin America, despite its incongruous name, has done excellent work in compiling information on military contracts and university involvement with the Pentagon. It has begun to develop techniques for making detailed maps of the military-industrial complex in local areas and attempting to instruct other groups in these research skills. Its address is P.O. Box 57, Cathedral Park Station, New York, N.Y. 10025.

Index

Richard J. Barnet

Richard J. Barnet is a recognized authority on problems of national security and arms control. A graduate of Harvard College and Harvard Law School, he studied foreign and military policy as a Fellow of the Harvard Russian Research Center after service in the U.S. Army as a legal officer specializing in problems of international law. During the Kennedy Administration he served in the State Department and the U.S. Arms Control and Disarmament Agency. In 1963 Mr. Barnet helped found the Institute for Policy Studies in Washington, an independent research and education center devoted to the study of public policy, and has since been its co-director. Mr. Barnet's previous books on foreign and military policy include *Who Wants Disarmament?* (1960), *After Twenty Years,* with Marcus Raskin (1965), and *Intervention and Revolution* (1968).

BOOKS OF POEMS BY MAY SWENSON

More Poems to Solve
Poems to Solve

Iconographs
Half Sun Half Sleep
To Mix with Time
A Cage of Spines
Another Animal

MORE POEMS
TO SOLVE

MAY SWENSON

CHARLES SCRIBNER'S SONS · NEW YORK

CONTENTS

Copy 1

PREFACE

A POEM IS A THING

A poem is a thing that can intensify the current of consciousness, make you see, hear, feel keenly—like an electrode to the brain. This power is not generated by the ideas so much as by the language in the poem. The poet works (and plays) with the elements of language, forming and transforming his material, to the point where a new perception emerges: something simple or ordinary may be seen as wonderful, something complex or opaque becomes suddenly clear.

A poet hopes that the output of the poem for the reader will equal the dynamic input—the initial "brain-touch"—that made him start the poem. But all expectations for its future, or any assessment of aims, occur only after the poem is done. It is a thing in itself, and it lets the writer know when it's done—*what* it has done—sometimes *why* it was done. Making it, the poet doesn't know, any more than you do until you read the last line, what it intends to show him. The making is a groping, a solving, a process something like trying to trap a flash of light into form.

Of the selections here, nine are Riddle Poems (so called because what they're about is not explicitly named) but they are planted as well with other hidden chances for surprise. In fact, be prepared to encounter something unexpected in every live poem.

Some of the Riddles are also Shapes, and these present a clue (even before you start to read) with their configuration on the page.

There are twelve Shaped Poems. Those not depicted by type

outlines extend their metaphors by more subtle visual means. Some run the opposite way of the Riddle. For instance, *Orbiter 5 Shows How Earth Looks From The Moon* states its subject whole in the title, and then goes on to symbolize it typographically. *Redundant Journey* quite literally *takes shape* on the page.

In six examples (some of which are Shaped Poems) certain words, letters within words, phrases or sentences are actualized, made into things physically there, by placement or other emphasis. Call them Word-Thing Poems. The moons in *Of Rounds* can be counted—visually, vocally—the space of the page, then, representing the boundary (as far as we can see) of our section of the universe. Repetitive rhythm adds to the poem's thingness. When it is read aloud the rotary pattern and movement of the spheres in their orbits are experienced. *How Everything Happens, Based on a Study of the Waves* makes the sentences concrete (liquid though they be!) by their placement, shift and movement. The words describe a process and simultaneously carry it out in a mimetic way. Another kind of acting-out of what is being shown takes place in *Fountains of Aix:* the many *waters* flowing down the righthand incline of the type-shape accumulate, while they delineate what the poem is about.

The serious play-poems in Section 3 most deliberately make use of word repetition, homonyms, or juggle with syntax in various ways. In *To Make a Play*, see how a 52-line poem is made, essentially, with only ten short key-words: *make, play, people, do, say, real, come, see, go, new*. Notice how their meanings shift with repetition and the changing context, and how the one word used only once is *new*. On your second reading, watch for the number of *watches* in *The Watch*. Solving *The Pregnant Dream*—keeping track of the quotes within the quotes—you may find, is like taking apart one of those painted wooden doll-boxes, with a replica in its belly, inside of which is another replica, and so on, down to the smallest belly and the last little doll. *MAsterMANANiMAl* is a mindful as well as a mouthful, since the language is in segments rather than sentences. With every M, A and N capitalized, it looks computer-made, almost. To experience extreme alliteration and assonance, read it aloud. Can you do it without getting twongue-tisted, and still gather all that it is saying?

Riddles, Shapes, Riddle-Shapes, Word-Things, Shape-Word-Things (not to mention the poems whose modes, by contrast, are more conventional) all present some unexpected facet or internal feature to be discovered *as part of* the recognition to be reached for in each. And there is yet another sort to solve here—at least one Riddle-Shape-Word-Thing. As its title suggests, *Camoufleur* (in

Section 5) is a paradigm of disguise, a hypnogogic interweave, both visual and verbal—confusing, at first, like a moire pattern. It seems to be for three eyes: your seeing eye, your reading eye and your mind's eye. The sequence and sense are camouflaged to tempt you to find them, by connecting the right juxtaposition of images. The Riddle is solved with identification of a certain bird, in its habitat, its appearance and behavior being factually yet metaphorically described. At the same time you may be led to take an insightful leap to a truth about human vanity.

If poetry is to remain art, it must not *merely* amuse, or mystify, or surprise. With the "shock of recognition" an empathic feedback to the reader's imagination ought to happen: a new apprehension of his own aliveness in the world—of his own potency for expression, as well as reception. In a way, human consciousness is one fabric. Our brains and bodies all have the same general shape and function, we share basic mental and sensual events. Poetry can magnify experience. Every poem is not alive for every person, but sometimes the act of solving, of locating its dynamics, may turn it on for you—and turn you on, too.

That a poem is *about* a thing is incidental to the fact that *it is a thing* in itself—a construct unlike any other built with language. Say the construct is its wires, say that what it's about is the particular vision illuminated when the power goes on—say *the power is the thing*. Its transmission quality will depend on its language, which the poet must keep rich and evocative but, at the same time, compact and exact. If all the elements are so connected and focused that the reader's whole consciousness is energized, feels a spontaneous "touch"—that charge is the first (and best) reward of the poem. Exploring the apparatus by which it operates can be an added exercise in pleasure.

May Swenson
1970

1
SPACE AND
FLIGHT POEMS

AFTER THE FLIGHT OF RANGER 7

Moon, old fossil,
to be scrubbed

and studied like
a turtle's stomach,

prodded over
on your back,

invulnerable hump
that stumped us,

pincers prepare to
pick your secrets,

bludgeons of light
to force your seams.

Old fossil, glistening
in the continuous rain

of meteorites
blown to you from

between the stars,
stilt feet mobilize

to alight upon you,
ticking feelers

determine your fissures,
to impact a pest

of electric eggs
in the cracks

of your cold
volcanoes. Tycho,

Copernicus, Kepler,
look for geysers,

strange abrasions,
zodiacal wounds.

13

ORBITER 5 SHOWS
HOW EARTH LOOKS FROM THE MOON

There's a woman in the earth, sitting on
her heels. You see her from the back, in three-
quarter profile. She has a flowing pigtail. She's
holding something
in her right hand—some holy jug. Her left arm is thinner,
in a gesture like a dancer. She's the Indian Ocean. Asia is
light swirling up out of her vessel. Her pigtail points to Europe
and her dancer's arm is the Suez Canal. She is a woman
in a square kimono,
bare feet tucked beneath the tip of Africa. Her tail of long hair is
the Arabian Peninsula.   A woman in the earth.

                                                                    A man in the moon.

Note: A telephoto of the earth, taken from above the moon by Lunar Orbiter
5 (printed in *The New York Times* August 14, 1967) appeared to show the
shadow-image of "a woman in a square Kimono" between the shapes of
the continents. The title is the headline over the photo.

FIRST WALK ON THE MOON

Ahead, the sun's face in a flaring hood,
was wearing the moon, a mask of shadow
that stood between. Cloudy earth
waned, gibbous, while our target grew:
an occult bloom, until it lay beneath
the fabricated insect we flew. Pitched
out of orbit we yawed in, to impact
softly on that circle.

                Not "ground"
the footpads found for traction.
So far, we haven't the name.
So call it "terrain," pitted and pocked
to the round horizon (which looked
too near): a slope of rubble where
protuberant cones, dish-shaped hollows,
great sockets glared, half blind
with shadow, and smaller sucked-in folds
squinted, like blowholes on a scape
of whales.

                Rigid and pneumatic, we
emerged, white twin uniforms on the dark
"mare," our heads transparent spheres,
the outer visors gold. The light was
glacier bright, our shadows long,
thin fissures, of "ink." We felt neither
hot nor cold.

                Our boot cleats sank
into "grit, something like glass,"
but sticky. Our tracks remain
on what was virgin "soil." But that's
not the name.

THREE JET PLANES

Three jet planes skip above the roofs
   through a tri-square of blue
   tatooed by TV crossbars
   that lean in cryptic concert in their wake

Like skaters on a lake
   combined to a perfect arrowhead up there
   they sever space with bloodless speed
   and are gone without a clue
   but a tiny bead the eye can scarcely find
   leaving behind
   where they first burst into blue
   the invisible boiling wind of sound

As horsemen used to do
As horsemen used to gallop through
   a hamlet on hunting morn
   and heads and arms were thrust
   through windows
   leaving behind them the torn
   shriek of the hound
   and their wrestling dust

Above the roofs three jet planes
   leave their hoofs of violence on naïve ground

OVER THE FIELD

They have
a certain
beauty, those
wheeled
fish, when over the field, steel fins stiff
out from
their sides,
they grope,

and then
through cloud
slice
silver snouts,
and climb,
trailing glamorous veils like slime.

Their long abdomens cannot curve, but
arrogant cut
blue, power
enflaming

their gills.
They claim
that sea where no fish swam until they flew
to minnow it
with their
metal.

The inflexible bellies carry, like roe,
Jonahs
sitting
row on row.
I sit by the
fin, in

one of those whale-big, wheeled fish, while
several silver
minnows line
up, rolling
the runway way
below.

WINDOW IN THE TAIL

Nap of cloud, as thick as stuffing
tight packed for a mattress ticking,

pickaninny kinked and puffed
and white as kid-sheared belly ruff,

is the floor and is the ceiling
over which we're keeled and sailing,

on flat pinion—not of feather—
but slatted aluminum or other

metal maneuverable
by ample ramps that bevel

up, or slide out wide
and glide

our carriage level.
Over fur of cloud we travel.

SLEEPING OVERNIGHT ON THE SHORE

Earth turns
        one cheek to the sun
while the other tips
        its crags and dimples into shadow.
We say sun comes up,
        goes down,
but it is our planet's incline
        on its shy invisible neck.
The smooth skin of the sea,
        the bearded buttes of the land
blush orange,
        we say it is day.
Then earth in its turning
        slips half of itself away
from the ever burning.
        Night's frown
smirches earth's face,
        by those hours marked older.
It is dark, we say.
        But night is a fiction
hollowed at the back of our ball,
        when from its obverse side
a cone of self-thrown shade
        evades the shining,
and black and gray
        the cinema of dreams streams through
our sandgrain skulls
        lit by our moon's outlining.

Intermittent moon
        that we say climbs
or sets, circles only.
        Earth flicks it past its shoulder.
It tugs at the teats of the sea.
        And sky

is neither high
        nor is earth low.
There is no dark
        but distance
between stars.
        No dawn,
for it is always day
        on Gas Mountain, on the sun—
and horizon's edge
        the frame of our eye.

Cool sand on which we lie
        and watch the gray waves
clasp, unclasp
        a restless froth of light,
silver saliva of the sucking moon—
        whose sun is earth
who's moon to the sun—
        To think this shore,
each lit grain plain
        in the foot-shaped concaves
heeled with shadow,
        is pock or pocket
on an aging pin
        that juggler sun once threw,
made twirl among
        those other blazing objects out
around its crown.
        And from that single toss
the Nine still tumble—
        swung in a carousel of staring light,
where each rides ringleted
        by its pebble-moons—
white lumps of light
        that are never to alight,
for there is no down.

MOON
  round
    goes around while going around a
          round
           EARTH

EARTH
  round
    with MOON
      round
        going around while going around
goes around while going around a
       round
        SUN

SUN
  round
    with EARTH
      round
       with MOON
         round
          going around while going
around, and MERCURY
      round
      and VENUS
        round
         going around while
going around, and MARS
      round
       with two MOONS
         round
         round
          going around
while going around, and JUPITER
      round
       with twelve MOONS
           round
           round
           round
           round
           round
           round
           round
           round
           round
           round
           round
           round 23

going around while going around, and SATURN
                                    round
                                       with nine
MOONS
        round
        round
        round
        round
        round
        round
        round
        round
        round
            going around while going around, and URANUS
                                            round
with five MOONS
            round
            round
            round
            round
            round
                going around while going around, and NEPTUNE
round
    with two MOONS
                round
                round
                    going around while going around, and
PLUTO
      round
            going around while going around, goes around while
going around
            A   OF ROUNDS
                                Round

2
WATER POEMS

AT TRURO

The sea is unfolding scrolls
and rolling them up again.
It is an ancient diary

the waves are murmuring.
The words are white curls,
great capitals are seen

on the wrinkled swells.
Repeated rhythmically
it seems to me I read

my own biography.
Once I was a sea bird.
With beak a sharp pen,

I drew my signature on air.
There is a chapter when,
a crab, I slowly scratched

my name on a sandy page,
and once, a coral, wrote
a record of my age

on the wall of a water-grotto.
When I was a sea worm
I never saw the sun,

but flowed, a salty germ,
in the bloodstream of the sea.
There I left an alphabet

but it grew dim to me.
Something caught me in its net,
took me from the deep

book of the ocean, weaned me,
put fin and wing to sleep,
made me stand and made me

face the sun's dry eye.
On the shore of intellect
I forgot how to fly

above the wave, below it.
When I touched my foot
to land's thick back,

it stuck like stem or root.
In brightness I lost track
of my underworld

of ultraviolet wisdom.
My fiery head furled
up its cool kingdom

and put night away.
The sea is unfolding scrolls,
and rolling them up.

As if the sun were blind
again I feel the suck
of the sea's dark mind.

OUT OF THE SEA, EARLY

A bloody
egg yolk. A burnt hole
spreading in a sheet. An en-
raged rose threatening to bloom.
A furnace hatchway opening, roaring.
A globular bladder filling with immense
juice. I start to scream. A red hydrocepha-
lic head is born, teetering on the stump of
its neck. When it separates, it leaks rasp-
berry from the horizon down the wide esca-
lator. The cold blue boiling waves cannot
scour out that band, that broadens, slid-
ing toward me up the wet sand slope. The
fox-hair grows, grows thicker on the
upfloating head. By six o'clock,
diffused to ordinary gold,
it exposes each silk thread and rumple in the carpet.

WAKING FROM A NAP ON THE BEACH

Sounds like big
rashers of bacon frying.
I look up from where I'm lying
expecting to see stripes

red and white. My eyes drop shut,
stunned by the sun.
Now the foam is flame, the long
troughs charcoal, but

still it chuckles and sizzles, it
burns and burns, it never gets done.
The sea is that
fat.

BEGINNING TO SQUALL

A Buoy like a man in a red sou'wester
is uP to the toP of its Boots in the water
    leaning to warn a Blue Boat

        that, BoBBing and shrugging, is nodding "No,"
            till a strong wave comes and it shivers "Yes."
                The white and the green Boats are quiBBling, too.
                What is it they don't want to do?

The Bay goes on Bouncing anchor floats,
their colors tennis and tangerine.
Two ruffled gulls laughing are laughing gulls,
    a finial Pair on the gray Pilings.

        Now the Boats are Buttoning slickers on
            which resemBle little tents.
                The Buoy is jumPing uP and down
                    showing a Black Belt stenciled "1."

A yellow Boat's last to lower sail
to wraP like a Bandage around the Boom.
    Blades are sharPening in the water
        that Brightens while the sky goes duller.

HOW EVERYTHING HAPPENS (Based on a Study of the Wave)

                                                    happen.
                                              to
                                        up
                                  stacking
                              is
                        something
When nothing is happening

When it happens
            something
                        pulls
                              back
                                    not
                                          to
                                                happen.

When   has happened.
      pulling back   stacking up
                  happens

            has happened   stacks up.
When it   something   nothing
                        pulls back while

Then nothing is happening.

                                          happens.
                                    and
                              forward
                        pushes
                  up
            stacks
      something
Then

31

3
WORD POEMS

TO MAKE A PLAY

To make a play
is to make people,
to make people do
what you say;

to make real people
do and say
what you make;
to make people make

what you say real;
to make real
people make up
and do what you

make up. What you
make makes people
come and see
what people do

and say, and then
go away and do
what they see—
and see what

they do. Real
people do and say,
and you see and
make up people;

people come to see
what you do.
They see what *they*
do, and they

may go away undone.
You can make
people, or you
can unmake. You

can do or you
can undo. People
you make up make up
and make people;

people come to
see—to see
themselves real,
and they go away

and do what you
say—as if they
were made up,
and wore make-up.

To make a play
is to make
people; to make
people make

themselves; to
make people
make themselves
new. So real.

When I
took my
watch to the watchfixer I
felt privileged but also pained to watch the operation. He
had long fingernails and a voluntary squint. He
fixed a magnifying cup over his
squint eye. He
undressed my
watch. I
watched him
split her
into three layers and lay her
middle (a quivering viscera) in a circle on a little plinth. He
shoved shirtsleeves up, and leaned like an ogre over my
naked watch, and with critical pincers poked and stirred. He
lifted out little private things with a magnet too tiny for me
to watch, almost. "Watch out!" I
almost said. His
eye watched, enlarged, the secrets of my
watch, and I
watched anxiously. Because what if he
touched her
ticker too rough, and she
gave up the ghost out of pure fright? Or put her
things back backwards so she'd
run backwards after this? Or he
might lose a minuscule part, connected to her
exquisite heart, and mix her
up, instead of fix her.
And all the time,
all the time
pieces on the walls, on the shelves told the time,
told the time
in swishes and ticks,
swishes and ticks,
and seemed to be gloating as they watched and told. I
felt faint, I
was about to lose my
breath (my
ticker going lickety-split) when watchfixer clipped her
three slices together with a gleam and two flicks of his
tools like chopsticks. He
spat out his
eye, and lifted her
high, gave her
a twist, set her
hands right, and laid her
little face, quite as usual, in its place on my
wrist.

38

ANALYSIS OF BASEBALL

It's about
the ball,
the bat,
and the mitt.
Ball hits
bat, or it
hits mitt.
Bat doesn't
hit ball, bat
meets it.
Ball bounces
off bat, flies
air, or thuds
ground (dud)
or it
fits mitt.

Bat waits
for ball
to mate.
Ball hates
to take bat's
bait. Ball
flirts, bat's
late, don't
keep the date.
Ball goes in
(thwack) to mitt,
and goes out
(thwack) back
to mitt.

Ball fits
mitt, but
not all
the time.
Sometimes
ball gets hit
(pow) when bat
meets it,
and sails
to a place
where mitt
has to quit
in disgrace.
That's about
the bases
loaded,
about 40,000
fans exploded.

It's about
the ball,
the bat,
the mitt,
the bases
and the fans.
It's done
on a diamond,
and for fun.
It's about
home, and it's
about run.

THE PREGNANT DREAM

I had a dream in which I had a
dream,
and in my dream I told you,
"Listen, I will tell you my
dream." And I began to tell you. And
you told me, "I haven't time to listen while you tell your
dream."

Then in my dream I
dreamed I began to
forget my
dream.
And I forgot my
dream.
And I began to tell you, "Listen, I have
forgot my
dream."
And now I tell you: "Listen while I tell you my
dream, a
dream
in which I dreamed I
forgot my
dream,"
and I begin to tell you: "In my dream you told me, 'I haven't time to
listen.'"

And you tell me: "You dreamed I wouldn't
listen to a
dream that you
forgot?
I haven't time to listen to
forgotten
dreams."
"But I haven't forgot I
dreamed," I tell you,
"a dream in which I told you,
'Listen, I have
forgot,' and you told me, 'I haven't time.'"
"I haven't time," you tell me.

And now I begin to forget that I
forgot what I began to tell you in my
dream.
And I tell you, "Listen,
listen, I begin to
forget."

40

MAsterMANANiMAl

ANiMAte MANANiMAl MAttress of Nerves
MANipulAtor Motor ANd Motive MAker
MAMMAliAN MAtrix MAt of rivers red
MortAl MANic Morsel Mover shAker

MAteriAl-MAster MAsticAtor oxygeN-eAter
MouNtAiN-MouNter MApper peNetrAtor
iN MoNster MetAl MANtle of the Air
MAssive wAter-surgeoN prestidigitAtor

MAchiNist MAsoN MesoN-Mixer MArble-heAver
coiNer cArver cities-idols-AtoMs-sMAsher
electric lever Metric AlcheMist
MeNtAl AMAzer igNorANt iNcubAtor

cANNibAl AutoMANANiMAl cAllous cAlculAtor
Milky MAgNetic MAN iNNoceNt iNNovAtor
MAlleAble MAMMAl MercuriAl ANd MAteriAl
MAsterANiMAl ANd ANiMA etheriAl

$$M = 55$$
$$A = 77$$
$$N = 42$$

4
COLOR AND
SOUND POEMS

THE BLINDMAN

The blindman placed
a tulip on his tongue for purple's taste.
Cheek to grass, his green

was rough excitement's sheen
of little whips.
In water to his lips

he named the sea blue and white,
the basin of his tears and fallen beads of sight.
He said: This scarf is red;

I feel the vectors to its thread
that dance down from the sun. I know
the seven fragrances of the rainbow.

I have caressed
the orange hair of flames. Pressed
to my ear,

a pomegranate lets me hear
crimson's flute.
Trumpets tell me yellow. Only ebony is mute.

FLAG OF SUMMER

Sky and sea and sand,
fabric of the day.
The eye compares each band.

Parallels of color on bare
canvas of time-by-the-sea.
Linen-clean the air.

Tan of the burlap
beach scuffed with prints
of bathers. Green and dapple,

the serpentine swipe
of the sea unraveling
a ragged crepe

on the shore. Heavy satin
far out, the coil,
darkening, flattens

to the sky's rim.
There a gauze screen,
saturate-blue, shimmers.

Blue and green and tan,
the fabric changes hues
by brush of light or rain:

sky's violet bar
leans over flinty waves
opaque as the shore's

opaline grains; sea silvers,
clouds fade to platinum,
the sand-mat ripples

with greenish tints
of snakeskin, or drying,
whitens to tent-cloth

spread in the sun. These bands,
primary in their dimensions,
elements, textures, strands:

the flag of summer,
emblem of ease, triple-striped,
each day salutes the swimmer.

COLORS WITHOUT OBJECTS

Colors without objects—colors alone—
wriggle in the tray of my eye,

incubated under the great flat lamp
of the sun:

bodiless blue, little razor-streak,
yellow melting like a firework petal,

double purple yo-yo
in a broth of murky gold.

Sharp green squints I have never seen
minnow-dive the instant they're alive;

bulb-reds with flickering cilia
dilate, but then implode

to discs of impish scotomata
that flee into the void;

weird orange slats of hot thought
about to make a basket—but

there is no material here—they slim
to a snow of needles, are erased.

Now a mottling takes place.
All colors fix chromosomic links

that dexterously mix,
flip, exchange their aerial ladders.

Such stunts of speed and metamorphosis
breed impermanent, objectless acts,

a thick, a brilliant bacteria—
but most do not survive.

I wait for a few iridium specks of idea
to thrive in the culture of my eye.

STONE GULLETS

Stone gullets among | Inrush | Feed | Backsuck and

The boulders swallow / Outburst | Hugh engorgements \ Swallow

In gulps the sea (Tide crams jagged | Smacks snorts chuckups) Follow

In urgent thirst | Jaws the hollow | Insurge (Hollow

Gushing evacuations follow | Jetty it must) Outpush | Greed

ELECTRONIC SOUND

A pebble swells to a boulder at low speed.
                    At 7½ *ips* a hiss is a hurricane.
                              The basin drain
is Charybdis sucking
                         a clipper down, the ship
                              a paperclip
whirling. Or gargle, brush your teeth, hear
                    a winded horse's esophagus lurch
                         on playback at 15/16. Perch
a quarter on edge on a plate, spin:
                    a locomotive's wheel is wrenched loose,
                wobbles down the line to slam the caboose,
keeps on snicking over the ties
                    till it teeters on the embankment,
                         bowls down a cement
ramp, meanders onto the turnpike
                              and into a junkhole
                    of scrapped cars. Ceasing to roll,
it shimmies, falters . . .
                         Sudden inertia causes
                              pause.
Then a round of echoes
                    descending, a minor yammer
as when a triangle's nicked by the slimmest hammer.

5
CREATURE POEMS

A PAIR

A he
and she,
prowed upstream,
soot-brown
necks,
bills the green
of spring
asparagus,

heads
proud figure-
heads for the boat-
bodies, smooth
hulls on feathered   the two,
water,   browed with light,
                            steer ashore,
                              rise; four
                                web-
                          paddles pigeon-
                                    toe it
                            to the reeds;

                                    he
                                walks first,
                              proud, prowed
                                as when light-
                              browed, swimming,
                                    he leads.

CAMOUFLEUR

Walked in the swamp   His cheek vermilion
         A dazzling prince
Neck-band white   Cape he trailed
       Metallic mottled
Over rain-rotted leaves   Wet mud reflected
       Waded olive water
His opulent gear   Pillars of the reeds
       Parted the strawgold
    Brilliance   Made him disappear

GEOMETRID

Writhes, rides down
on his own spit,
lets breeze twist

him so he chins,
humps, reels up it,
munching back

the vomit string.
Some drools
round his neck.

Arched into a staple
now, high on green
oak leaf he punctures

for food, what
was the point
of his act? Not

to spangle the air,
or show me his trick.
Breeze broke

his suck,
so he spit
a fraction of self's

length forth, bled
colorless from within,
to catch a balance,

glide to a knot
made with his own mouth.
Ruminant

while climbing, got
back better than bitten
leaf. Breeze

that threw
him snagged him
to a new.

CATBIRD IN REDBUD

Catbird in the redbud this morning.
No cat could
mimic that rackety cadenza he's making.
And it's not red,
the trapeze he's swaying on.
After last night's freeze,
redbud's violet-pink, twinkled on
by the sun. That bird's
red, though, under the tail
he wags, up sharply, like a wren.

The uncut lawn hides blue
violets with star-gold eyes on the longest
stems I've ever seen. Going to
empty the garbage, I simply have
to pick some,
reaching to the root of green,
getting my fist dewy, happening
to tear up a dandelion, too.

Lilac, hazy blue—
violet, nods buds over the alley
fence, and (like a horse with a yen
for something fresh for breakfast)
I put my nose into a fragrant
pompom, bite off some, and chew.

Unconscious
came a beauty to my
wrist
and stopped my pencil,
merged its shadow profile with
my hand's ghost
on the page:
Red Spotted Purple or else Mourning
Cloak,
paired thin as paper wings, near black,
were edged on the seam side poppy orange,
as were its spots.

UNCONSCIOUS
CAME A BEAUTY

I sat arrested, for its soot haired
body's worm
shone in the sun.
It bent its tongue long as
a leg
black on my skin
and clung without my
feeling,
while its tomb stained
duplicate parts of
a window opened.
And then I
moved.

Copy 1

58

REDUNDANT JOURNEY

I'll rest here in the bend of my tail
said the python having traveled
his own length
beginning with his squared snout
laid beside his neck
O where does the neck
end and the chest begin
O where does the stomach
end and the loins begin
O where are the arms and legs
Now I'll travel between myself
said the python lifting his snout
and his blue eyes saw lead-gray
frames like windows on his hide
the glisten of himself the chill
pattern on each side
of himself and as his head slept
between the middles of himself
the end of his outer self still crept
The python reared his neck and yawned
his tongue was twins his mucous membrane
purple pink hibiscus sticky
He came to a cul de sac in the lane
of the center of his length
his low snout
trapped between twin windowed
creeping hills of himself
and no way out
I'll travel upon myself said the python
lifting his chin to a hill
of his inner length and while
his neck crossed one half of his
stomach his chest crossed his
loins while his tail lay still
But then he thought
I feel uncomfortable in
this upright knot
and he lowered his chin
from the shelf of himself
and tucked his snout in
How get away from myself said
the python beside himself
traveling his own side
How recognize myself as just myself
instead of a labyrinth I must travel
over and over stupified
His snout came to the end
of himself again to the final leaden bend
of himself
Said the python to his tail
Let's both rest till all
the double windowed middle maze
of ourself
gets through crawling

MOTHERHOOD

She sat on a shelf,
her breasts two bellies
on her poked-out belly,
on which the navel looked
like a sucked-in mouth—
her knees bent and apart,
her long left arm raised,
with the large hand knuckled
to a bar in the ceiling—
her right hand clamping
the skinny infant to her chest—
its round, pale, new,
soft muzzle hunting
in the brown hair for a nipple,
its splayed, tiny hand picking
at her naked, dirty ear.
Twisting its little neck,
with tortured, ecstatic eyes
the size of lentils, it looked
into her severe, close-set,
solemn eyes, that beneath bald
eyelids glared—dull lights
in sockets of leather.

She twitched some chin-hairs,
with pain or pleasure,
as the baby-mouth found and
yanked at her nipple;
its pink-nailed, jointless
fingers, wandering her face,
tangled in the tufts
of her cliffy brows.
She brought her big
hand down from the bar—
with pretended exasperation
unfastened the little hand,
and locked it within her palm—
while her right hand,
with snag-nailed forfinger
and short, sharp thumb, raked
the new orange hair

of the infant's skinny flank—
and found a louse,
which she lipped, and
thoughtfully crisped
between broad teeth.
She wrinkled appreciative
nostrils which, without a nose,
stood open—damp, holes
above the poke of her mouth.

She licked her lips, flicked
her leather eyelids—
then, suddenly flung
up both arms and grabbed
the bars overhead.
The baby's scrabbly fingers
instantly caught the hair—
as if there were metal rings there—
in her long, stretched armpits.
And, as she stately swung,
and then proudly, more swiftly
slung herself from corner
to corner of her cell—
arms longer than her round
body, short knees bent—
her little wild-haired,
poke-mouthed infant hung,
like some sort of trophy,
or decoration, or shaggy medal—
shaped like herself—but new,
clean, soft and shining
on her chest.

A BIRD'S LIFE

Is every day a separate life to a bird?
Else why,
as dawn finds the slit lid of starling- or sparrow-eye,
spurts that mad bouquet from agape bills?
Streamered, corkscrew, soprano tendrils
riot in the garden—
incredulous ejaculations at the first pinches
of birth. Tiny winches
are tightened, then hysterically jerked loose.
There is produced
a bright geyser of metal-petaled sound
that, shredding, rubs its filings into my sleep.

As the sun, Herself, bulges from a crack in the cloud-shell,
a clamp is applied to every peep—
a paralysis of awe, as they are ovened
under the feathers of Mother Light—
a stun
of silence. Then they revert
to usual.

When the sun
is higher, only a blurt
of chitters, here and there,
from the sparrows—
sassy whistles, sarcastic barks from the starlings.
By noon they're into middle age
and the stodgy business of generation.

Evening, though, leaks
elegy from a few pathetic beaks.
Chirks of single-syllable despair
that the sky is empty and their
flit-lives almost done.
Their death is the death of light.
Do they lack memory, and so
not know
that the Hen
of the sun
will hatch them again
next morning?

NEWS FROM THE CABIN

1

Hairy was here.
He hung on a sumac seed pod.
Part of his double tail hugged the crimson
       scrotum under cockscomb leaves—
         or call it blushing lobster claw, that swatch—
       a toothy match to Hairy's red skullpatch.
Cried *peek!* Beaked it—chiselled the drupe.
His nostril I saw, slit in a slate whistle.
White-black dominoes clicked in his wings.
Bunched beneath the dangle he heckled with holes,
       bellysack soft, eye a brad, a red-flecked
       mallet his ball-peen head, his neck its haft.

2

Scurry was here.
He sat up like a six-inch bear,
       rocked on the porch with me;
       brought his own chair, his chow-haired tail.
Ate a cherry I threw.
Furry paunch, birchbark-snowy, pinecone-brown back
       a jacket with sleeves to the digits.
Sat put, pert, neat, in his suit and his seat, for a minute,
       a frown between snub ears—bulb-eyed head
       toward me sideways, chewed.
Rocked, squeaked. Stored the stone in his cheek.
Finished, fell to all fours, a little roan couch;
       flurried paws loped him off, prone-bodied,
       tail turned torch, sail, scarf.

~~J821~~
MS   811.54   Swenson, May   Copy 1

More poems to
solve

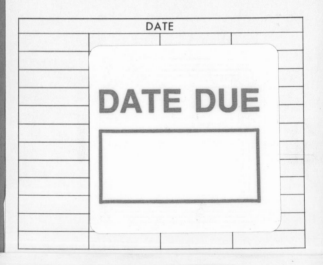

DATE

DATE DUE